Agent-Based Social Systems
Volume 3

Series Editor: Hiroshi Deguchi, Yokohama, Japan

ABSS — Agent-Based Social Systems

This series is intended to further the creation of the science of agent-based social systems, a field that is establishing itself as a transdisciplinary and cross- cultural science. The series will cover a broad spectrum of sciences, such as social systems theory, sociology, business administration, management information science, organization science, computational mathematical organization theory, economics, evolutionary economics, international political science, jurisprudence, policy science, socioinformation studies, cognitive science, artificial intelligence, complex adaptive systems theory, philosophy of science, and other related disciplines.

The series will provide a systematic study of the various new cross-cultural arenas of the human sciences. Such an approach has been successfully tried several times in the history of the modern sciences of humanities and systems and has helped to create such important conceptual frameworks and theories as cybernetics, synergetics, general systems theory, cognitive science, and complex adaptive systems.

We want to create a conceptual framework and design theory for socioeconomic systems of the twenty-first century in a cross-cultural and transdisciplinary context. For this purpose we plan to take an agent-based approach. Developed over the last decade, agent-based modeling is a new trend within the social sciences and is a child of the modern sciences of humanities and systems. In this series the term "agent-based" is used across a broad spectrum that includes not only the classical usage of the normative and rational agent but also an interpretive and subjective agent. We seek the antinomy of the macro and micro, subjective and rational, functional and structural, bottom-up and top-down, global and local, and structure and agency within the social sciences. Agent-based modeling includes both sides of these opposites. "Agent" is our grounding for modeling; simulation, theory, and real-world grounding are also required.

As an approach, agent-based simulation is an important tool for the new experimental fields of the social sciences; it can be used to provide explanations and decision support for real-world problems, and its theories include both conceptual and mathematical ones. A conceptual approach is vital for creating new frameworks of the worldview, and the mathematical approach is essential to clarify the logical structure of any new framework or model. Exploration of several different ways of real-world grounding is required for this approach. Other issues to be considered in the series include the systems design of this century's global and local socioeconomic systems.

T. Terano, H. Kita, H. Deguchi,
K. Kijima (Eds.)

Agent-Based Approaches in Economic and Social Complex Systems IV

Post-Proceedings of The AESCS International Workshop 2005

With 99 Figures

Springer

Takao Terano, Ph.D.
Professor, Interdisciplinary Graduate School of Science and Engineering
Tokyo Institute of Technology
4259 Nagatsuta-cho, Midori-ku, Yokohama 226-8502, Japan

Hajime Kita, Dr. Eng.
Professor, Academic Center for Computing and Media Studies
Kyoto University
Yoshida-Nihonmatsu-cho, Sakyo-ku, Kyoto 606-8501, Japan

Hiroshi Deguchi, Ph.D.
Professor, Interdisciplinary Graduate School of Science and Engineering
Tokyo Institute of Technology
4259 Nagatsuta-cho, Midori-ku, Yokohama 226-8502, Japan

Kyoichi Kijima, Dr. Eng.
Professor, Department of Value and Decision Science
Tokyo Institute of Technology
2-12-1 O-okayama, Meguro-ku, Tokyo 158-8550, Japan

Library of Congress Control Number: 2007923065

ISSN 1861-0803
ISBN 978-4-431-71306-7 Springer Tokyo Berlin Heidelberg New York

Springer is a part of Springer Science+Business Media
springer.com

Typesetting: Camera-ready by the editors and authors

Printed on acid-free paper

Preface

As interests in agent-based modeling in economic and social systems spreads more widely, today, we can find various meetings, proceedings, and journals on these topics. Among them, we of the Pacific-Asian Association for Agent-Based Approaches (PAAA) in social system sciences have hosted a series of workshops: Agent-Based Approaches in Economic and Social Complex Systems (AESCS). The objective of the AESCS workshops is to promote the worldwide activities of the multidisciplinary community of multiagent, computational economics, organizational science, social dynamics, and complex systems. AESCS workshops bring together researchers and practitioners from diverse fields, such as computer science, economics, physics, sociology, psychology, and complex theory, in order to understand emergent and collective phenomena in economic, organizational, and social systems. The first, second, and third workshops were held, respectively, in Shimane City, on May 21-22, 2001, in conjunction with the annual conference of the Japanese Society for Artificial Intelligence; at Tokyo University on August 16 2002; and at Kyoto University May 27-29, 2004.

The Fourth International Workshop on Agent-Based Approaches in Economic and Social Complex Systems (AESCS 2005), "Creation of Agent-Based Social System Sciences", was held at the Tokyo Institute of Technology July 9 - 13, 2005, with the support of the 21st COE Program.* We had more than sixty submissions from several countries. Among them, 47 papers were presented at AESCS 2005. All papers were reviewed by at least two reviewers in the domain. The reviewers' comments were carefully examined and discussed by the program committee members to ensure fairness and consistency in the selection process. As a result of those serious discussions, in this volume we have presented 22 selected papers from the presentations—a final acceptance rate of approximately 40% of submissions.

The papers included are categorized into the following subtopics, which cover the broad contemporary areas of agent-based modeling theories, methodologies, and applications: Gaming and Education (3 papers), Market and Policy (3 papers), City and Local Community (3 papers), Networks (5 papers), Learning Agents (3 papers), and Methodology and Tools (5 papers). We believe that this volume will be highly attractive and useful for those who have an interest in the recent progress in agent-based simulation in the social sciences.

Finally, we would like to acknowledge the support and encouragement of many people in helping us to hold this workshop and to prepare this volume. We also thank Springer Japan for publishing the volume.

Takao Terano, General Chair
Organizing Chair, Kyoichi Kijima

---

*The 21st Century COE Program is an initiative by the Japanese Ministry of Education, Culture, Sports, Technology and Science (MEXT) to cultivate a competitive academic environment among Japanese universities by giving targeted support to the creation of world-standard research and education bases (Centers of Excellence) in a range of disciplines.

Conference Committee

Hiroshi Deguchi, Tokyo Institute of Technology, Japan
Kyoichi Kijima, Tokyo Institute of Technology, Japan
Takao Terano, Tokyo Institute of Technology, Japan
Yuhsuke Koyama, Tokyo Institute of Technology, Japan
Yusuke Arai, Tokyo Institute of Technology, Japan

Contents

Networks

Learning Agents

Methodology and Tools

Gaming and Education

Using U-Mart System for the lecture of economics*

Yuhsuke Koyama[1], Ko Ishiyama[1], Hiroyuki Kaneko[2], Isao Ono[1], and Hiroyuki Matsui[3]

[1] Interdisciplinary Graduate School of Science and Engineering, Tokyo Institute of Technology, 4259 Nagatsuta, Midori, Yokohama 226-8502, Japan, koyama@dis.titech.ac.jp, ishiyama@cs.dis.titech.ac.jp, isao@dis.titech.ac.jp
[2] Graduate School of Social Systems Science, Chiba Institute of Technology, 2-17-1 Tsudanuma, Narashino, Chiba, 275-0016, Japan, kaneko@rynx.com
[3] Graduate School of Economics, Kyoto University, Yoshida-Honmachi, Sakyo-ku, Kyoto 606-8501, Japan, hmatsui@econ.kyoto-u.ac.jp

Summary. We report the result of lectures using U-Mart which was held held at Chiba Institute of Technology, February 28 - March 2, 2005. We verified that without the orders from machine agents, orders from 150 students have enough variety to determine the market price. However, it is necessary to overcome various problems, especially incentive problems, to use data from lecture for research.

Key words. U-Mart, Gaming Simulation, Artificial Market ,education

1 Introduction

1.1 An outline of the U-Mart project and U-Mart System

The U-Mart(http://www.u-mart.econ.kyoto-u.ac.jp/) is an artificial futures market system, where human agents and machine agents can simultaneously participate in trading via LAN and the Internet. U-Mart is a simulator for the stock price index futures and time sereis of prices in J30, which is taken the average of the stock prices of Japan Top 30 enterprises, is used as the spot price.

The U-Mart system is a client-server system which uses a dedicated protocol built on TCP/IP to exchange such information as trading information on the Internet. The server that simulates the stock exchange accepts orders from clients, executes pricing and trading, and manages the asset account.

Each client obtains such information as price movements from the server and places orders based on their own decision. The form of the client is not a problem as long as it behaves according to the trading protocol. It means that aside from trading by program (machine agent), gaming simulation experiment by trading by human (human agents) only, as well as experiments where both machine and human agents are involved in trading are taken into consideration in designing the system.

* We received assistance from various participants in the U-Mart project in preparing this paper. This research was partially supported by the Ministry of Education, Science, Sports and Culture, Grant-in-Aid for Scientific Research on Priority Areas, 16016274, 2001-2004.

Both U-Mart server and client can work in Windows environment that can run JAVA applications. Every kind of result in simulation is saved as the series of text files in CSV format, so that they can be analyzed on a spreadsheet program and even social science faculty members without special computer skills can handle.

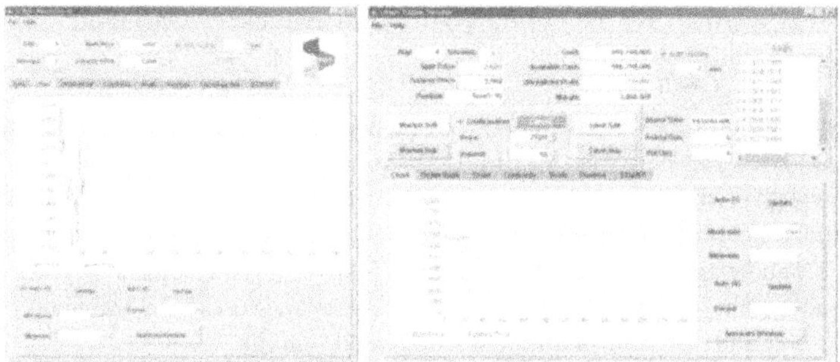

Fig. 1. Snapshot of U-Mart Server and Client

To tackle this complicated challenge, researchers from various fields, including engineering, economics and psychology, take part in and approach the problem from various fields, such as artificial intelligence, artificial market, cognitive science, and learning theory, in addition to conventional market study.

To promote such interdisciplinary study, it is necessary for reseachers from various fields to share a common ground Therefore, we think that it is important to share not only the subject of the study, namely designing the financial market, but also the test bed for the approach, namely the equipment necessary for the study. The U-Mart project was organized to provide a common test bed which can be shared by researchers who are interested in behavior of the financial market and other socioeconomics systems as well as behavior of economic bodies that operate there.

1.2 Educational Activities using U-Mart

A salient feature of the U-Mart project is that it associates three activity areas, namely research, events such as open experiments, and education to each other in order to understand and the market and to establish a control methodology. Accordingly, human experiments in college education are not only expected of producing educational effect but also occupy important position in the project, so we also eagerly collect know-how about performing the experiment and develop it as courseware.

The U-Mart system is used as courseware in college education in the fields of engineering and economics. At educational institutions in engineering field, the U-Mart system is used as assignment of programming exercise. In the field of economics, the system is used in various forms, for example, as a tool to practically understand the futures market as well as a teaching material of data mining using spreadsheet software.

These three activities are deeply interrelated. Machine agents collected at the open experiments are necessary to construct the fruitful agent set which are necessary to create the market-like condition where researchers evaluate the performance of their developed agents. Many of machine agents are developed by the students of U-Mart lecture. Various tools developed for the research are also used for education and events. Accumulation of open experiments not only contributes to log analysis but also identifies problems to be solved by artificial markets. Furthermore, with progress of research, the purpose of open experiments becomes clearer and rules and systems have been changed.

2 U-Mart Intensive Course at Chiba Institute of Technology in 2005

2.1 Required knowledge and skills and teaching materials

Gaming simulation or role playing is very useful for lecturers to make students understood the complex system through experiencing the game which simulates and represents the complex real world. U-Mart is very effective educational tool for students to understand the futures market system.

However, students have to study a lot of essential knowledge before starting the game. If we start U-Mart trading game after students completely understand, students have to study all following knowledge and skills.

1. Futures market
    - market institution
        - negotiated transaction, transaction in the exchange
    - price determination mechanism
        - *itayose*, *zaraba*(or double auction)
    - deposit money system and bunkruptcy
    - marking to the market
    - social function of the futures market
        - risk hedge , price expectation, liquidity supply
    - participants of the futures market
        - headger, arbitrager, specurator
2. Stock Price Index
    - Stock Price Index
    - J30
    - Other price index
        - TOPIX, Nikkei225, Nikkei 300
3. Trading strategy
    - basic trading method
        - trend and moving average, ordinary and contrary trading, arbitarage
    - position adjustment and risk controle
    - other technical analysis
4. Simulation
    - Gaming Simulation
        - concept, social experiment, institutional design
    - Agent Base Simulation

    - concept, Gaming Simulation and Agent Simulation, Human agent and Software Agent
  - Artificial Market
5. Rules of U-Mart experiment
  - initial assets
  - time structure of the experiment
    - 30 days, each day contains 8 *itayoses*
    - After the last (eigth) *itayose* is over, the assets of all clients are marked to the market.
  - deposit money system and bunkruptcy
  - (If necessary) strategies of machine agents to join the experimet
6. Operating skills for the GUI Client
  - order , cancel and contracted order list
  - price chart and board information
  - position chart

As it takes at least four or five hours to explain these knowledge, we cannot anticipate that students fully master the knowledge of the futures market and skill. So, we have to design the lecture plan for students to study futures market and U-Mart gradually.

2.2 Course materials

To fulfill these requirements, we prepared materials most of which are made of PowerPoint files and distributed to students in the format of PDF.

1. Introduction and Start Up
  - Introduction to lecture and virus elimination (I1)
  - Outline of the U-Mart Project (I2)
  - U-Mart and Software development(I3)
  - Start Up to U-Mart(I4)
2. Futures Market
  - Function of the futures market (F1)
  - Participants in the futures market (F2)
  - Mechanism of the futures market (F3)
  - Futures of the Stock index (F4)
  - Price determination algorithm (F5)
  - Trade tips(F6)
3. Introduction to Agent Based Modeling
  - Introduction to Agent Based Modeling(A1)
  - Interpretation of machine agents: about the standard agent kit(A2)
4. Instructions of the experiment
  - How to use GUI client (GUI1)
  - How to read Charts (GUI2)
5. Log data analysis
  - (Software) Log Analizer(L1)
  - How analyze the log data to use log analizer (L2)
  - Report of the second day's network experiment (L3)
6. Exercises
  - Exercises to calculate the deposit money and marking to market (g1)

- Excel files to explain the answer of the exercises (g2)

Though most of materials were prepared in advance of the U-Mart course, the report of the second day's experiment was necessary to make after the lecture of the second day. Although it is difficult to prepare the report of the network experience, it is very effective for students to understand the causal process of theri action to the price fluctuation.

2.3 U-Mart course

U-Mart Course was held at Faculty of Social System Science, Division of Project Management, Chiba institute of Technology, from February 28 to March 2. Participants to the lectures were approx. 160 students, which include students who took the class for the second time and students from other departments. In this course, because of lecture room restrictions, students are divided into two classrooms. Capacity of each class room is nearly 100.

As all students take laptop PC, computer room was not necessary. However, because of the lack of the conscousness for the network security, we had to begin the virus check to the all students laptops before starting U-Mart lectures and experiments.

Table 1. Outline of the U-Mart Course

Title	Special Issues in Project Management
Date	2/28/2005-3/2/2005
Faculty	Social System Science
Division	Project Management
Grade	Sophomore (2nd Grade)
Place	Bldg.9, room 7301,7302
Lecturer	7301: Matsui , 7302: Koyama
Teaching Assistant	Ko Ishiyama and Hiroyuki Kaneko
N of Students	160(80 in each classroom)
Style	lecture, standalone experiment and network experiment (in the classroom and all students)
Necessity	Laptop PC (for all students)

2.4 Outline of U-Mart Course and Required Resources

Through our experience, we reccomend the required resources for the lecture. They are as follows:

- Desktop PCs :Pentium 4(or faster) - 2GHz $\times$ 2
- (Strongly Recommend) USB Memories :128MB(or larger) $\times$ 20
- Teaching Assisntants :2 (or more)
- Room: Network socket available
- Others :projector, laptop PC for lecturer (each classroom)

Desktop PCs are used as the market server. As the deta recording speed is faster, we reccomend the desktop PCs other than laptops. 2GHz is enough CPU power to experiment with 100 students. If 150 students experiment is needed, 2.4GHz or faster is enough.

USB Memories is used to distribute the U-Mart Applications to students, collect the reports from students, and distribute the virus checker program. We strongly reccomend to utilize USB memories.

As some students' laptops may be infected with worms, toroys, and so on at the first lecture, we cannot the ftp or http server to distriute the U-Mart Clients via LAN. At the first lecture, we distributed the virus checker program via usb memories and made all students check their own laptops. In our lecture, about 15% of laptops were infected by worm Lonsan.*,Nachi.*,Sasser.*, or Korgo.*, toroy Win32/psw or PasorotK, and so on.

3 Results

3.1 Progress in Price Chart

To compare price chart is very easy way to assess the effect of training for students. The Left figure in Fig.2 represents the result of first experiments and the right one in Fig.2 represents the result of second day afternoon experiments. In the former, the shape of price fluctuation is not market-alike. There are a lot of huge fluctuations. On the other hand, in the latter, the shape of price fluctuation, without the last confusion periods, is closed to those of real market at a glance.

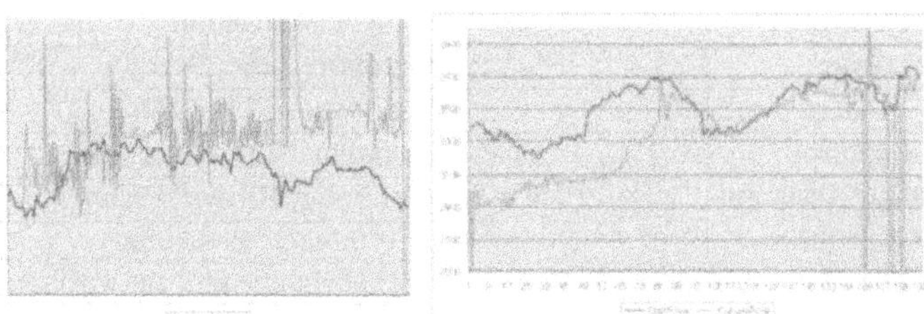

Fig. 2. Result in the 1st day morning and the 2nd day afternoon ,Room 7302

Fig.3 gives the price distribution of ΔP, which is defined of the difference of current price and last price. The Left figure in Fig.**??** shows the result of first experiments and the right one in Fig.2 shows the result of second day afternoon experiments. At the firstday, students' behaved extremely and the price distribution of futures prices is far from that of spot prices. However, at the second day, students' behavior improved drastically. Even though price distribution of futures prices shows stronger tendency of the high peak and fat tail than that of spot prices, they are very alike.

These figures and some students reports suggests that students as a group learned about the market behavior.

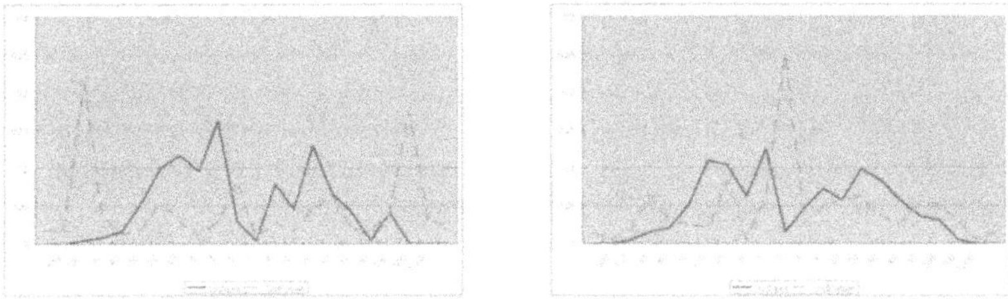

Fig. 3. Distribution of ΔP

3.2 Weak points of the large scale network gaming

U-Mart market server is robust enough to be connected with more than 100 students. GUI trading client, developed as the subset of the server, is also robust. However, for the success of the network gaming , we have to control a lot of human factors, especially keeping playing incentives.

In U-Mart gaming, all game players connect the market server and buy / sell the futures. As futures price is determined by the order from machine agents and human players, if someone send the huge amount order just for pleasure, it causes the price spike (sudden rise or fall). If this price is synchronized at the time of marking to market, some players are driven into the bankruptcy. Though such game is a lot of fun and students are much excited, it has no value for researcher to analyze the log data. One of the reasons for students to lose incentives is that their understanding is not enough and the U-Mart is too complex to keep their curiosity, i.e. they feel U-Mart gaming as the monotonous work. We have to design the course of study carefully to keep their curiosity and incentive to play.

4 Conclusion

Gaming simulation itself is a good educational tool for teaching the mechanism of the complex economic phenomena. However, if the target phenomena is very complicated, the game simplified from the real world is also very complicated for students to understand at a stretch. Elaborate course design is indispensable to succeed the lecture. We took the strategy to bundle the lecture, gaming and report writing. At first, we delivered the minimum knowledge and skill to play U-Mart gaming and started the U-Mart game. After students took a grasp of the outline of U-Mart system, report writing was assigned to students to digest their gaming experience. By repeating this cycle, we carried the required knowledge gradually and students gained the full understanding of the futures market. One of the surprising was the emergence of the speculator students who tried to control the market price fluctuation. We think it was the evidence that students had the deep understanding about the market mechanism. Though our plan is not matured well, it is well defined enough for the lecturers to practice at their class.

Thanks to improvements through numerous experiments in the past, the system is very easy-to-use and valuable as educational courseware. Especially this system is very handy as a platform of gaming simulation with the scalability that can be used in large-scale experiments of over 100 people and usage in various experiments. In addition, the U-Mart continuously grows as a system that can serve as a test bed for education and research. I wish many people will take interest and take part in this project and the system will be used even more actively.

References

1. H. Sato, Y. Koyama, K. Kurumatani, Y. Shiozawa, and H. Deguchi: "U-Mart: A Test Bed for Interdisciplinary Research in Agent Based Artificial Market", Evolutionary Controversies in Economics, 179-190, 2001.
2. T. Terano, Y. Shiozawa, H. Deguchi, H. Kita, H. Matsui, H. Sato, and I. Ono: "U-Mart: An Artificial Market to Bridge the Studies on Economics and Multiagent Systems", Proceedings of Fourth Pacific Rim International Workshop on Multi-agents, 371-385, 2001.
3. H. Matsui, I. Ono, H. Sato, H. Deguchi, T. Terano, H. Kita, and Y. Shiozawa: "Learning Economics Principles from the Bottom by both Human and Software Agents-Outline of U-Mart Project", -Proceedings of CASOS 2001 International Conference, 97-99, 2001.
4. H. Sato, H. Matsui, I. Ono, H. Kita, and T. Terano:"U-Mart Project: Learning Economic Principles from the Bottom by both Human and Software Agents",Proceedings of AESCS2001, 56-66, 2001.
5. H. Sato, H. Matsui, I. Ono, H. Kita, T. Terano, H.Deguchi, and Y. Shiozawa: "U-Mart Project: Learning Economic Principles from the bottom by Both Human and Software Agents", New Frontiers in Artificial Intelligence , 121-131, 2002.
6. H. Sato, H. Matsui, I. Ono, H. Kita, T. Terano, H. Deguchi, and Y. Shiozawa :"Case Report on U-Mart Experimental System: Competition of Software Agent and Gaming Simulation with Human Agents", Agent-Based Approaches in Economics and Social Complex Systems, Joint JSAI 2001 Workshop Pst-Proceedings, 167-178, 2002.
7. T. Terano, Y. Shiozawa, H. Deguchi, H. Kita, H. Matsui, H. Sato, I. Ono and Y. Nakajima: "U-Mart: An Artificial Market Testbed for Economics and Multiagent Systems", 2nd International Workshop on Agent-based Approaches in Economics and Social Complex Systems, 55-62, 2002.
8. K.Taniguchi, Y.Nakajima, and F.Hashimoto, "A report of U-Mart Experiments by Human Agents", R.Shiratori, K.Arai, F.Kato(Eds.) Gaming, Simulations, and Society Research Scope and Perspective, Springer-Verlag Tokyo, pp.49-57, 2004.
9. H. Matsui, Y. Koyama, K. Ishiyama: "A Report of Large-Scale Gaming Simulation using a U-Mart System in Economic Education", The Third International Conference on Creating, Connectiong, and Collaboration through Computing(C5), 179-184,2004.

Multi-Agent Participatory Simulations Between Experimental Economics and Role-Playing Games

Paul Guyot[1,2], **Alexis Drogoul**[3] and **Shinichi Honiden**[2]

[1]LIP6, Université Pierre et Marie Curie, paul.guyot@lip6.fr
[2]National Institute of Informatics, {paul,honiden}@nii.ac.jp
[3]IRD Bondy, alexis.drogoul@bondy.ird.fr

Abstract. In this paper, we present participatory experiments in economics and role-playing games from an agent-based simulations perspective. In particular, we try to compare these two traditions of participatory approaches to stress identities and differences. After having outlined the lessons of these traditions, we describe multi-agent participatory simulations as a new kind of participatory experiments combining agents and participation more closely, using as an example the methodology we developed to conduct experiments of this new kind.

Multi-agent participatory simulations are described as simulations where agents and players have similar roles and we propose to use assistant agents within these simulations for the purpose of participatory design of protocols or multi-agent systems.

Introduction

Economical experimentation and role-playing games represent two opposite ends on the spectrum of participatory experiments.

On one hand, economics probably has the highest concern for positiveness in the methodology chosen to conduct these experiments. Experiments in laboratories of economics are considered as equivalent to experiments in physics, with the idea that they can serve to test theories in a Popperian approach.

On the other hand, role-playing games are often used for their impact on participants and belong to the participatory action research tradition (Wadsworth, 1998). Recent work coupled role-playing games with multi-agent simulations in an iterative manner, the outcomes of the game being used to improve the multi-agent model it is based on. Such experiments belong to the companion modeling (ComMod) approach (Barreteau et al., 2003).

It might be important to stress that by role-playing games, we mean scientifically conducted participatory experiments where participants have a specific role, designed in the game scenario. The word « game » itself is confusing as it has a large range of acceptions in the various communities mentioned here: economists would link it with game theory, for example, even if experimental economics do not necessarily deal with game theory.

This paper is divided in three sections. In the first section, we will present experimental economics for multi-agent specialists and we will especially outline what can be learned from fifty years of experimentations in economics. In the second section, we will present role-playing games as they are used by scientists and especially their coupling with multi-agent systems. Finally, drawing from examples of experiments we conducted, we will argue in favor of a stronger coupling of role-playing games with multi-agent simulations in an approach that could be called multi-agent participatory simulations.

1 Lessons from Experimental Economics

Interest in experimental economics follows a growing trend. Several economists did conduct participatory experiments in the 1950s, and the founding paper for experimental economics as a method and as a discipline within economics is probably « Towards Experimental Economics » (Castro and Weingarten, 1970). Experimental economics received additional attention from mainstream economists since Vernon Smith was awarded the Nobel Prize in economics in 2002 (with Daniel Kahneman).

1.1 A finer description of « The Model »

The vocabulary used in economics and the one used within the multi-agent and agent-based simulation community are very different. Several terms overlap. Economics being centered on production and resources, the particularity of « agents » in economics is that they represent entities that communicate and exchange products and resources.

The notion of « model » is quite vague within the multi-agent simulation community. Probably because « model » has a very strict acception in economics that is not linked with experiments, economists have developed four key concepts that replace our « model » (after Smith, 1994).

The **environment** is what we would call the properties of agents. Since economists agents are entities that only communicate, produce and exchange resources and products, these properties include the initial amounts of resources they have, their tastes and the production function, i.e. how they transform resources with their labour.

The **institution** is the mode of interaction between agents. It mostly consists in rules determining how agents can behave and communicate together. In an experiment based on auctions, the way the auctions are conducted is part of the institution.

The **behavior** is what the agents chooses to do considering his properties (the environment) and the practices (the institutional rules). Interestingly, this distinction doesn't follow the data and code distinction, since the production or the preference functions would typically be coded.

Finally, the **outcomes** are the final resource allocations after the experiment as they can be observed by the experimenters.

What interest experimental economists is to study what they call the **behavior**. This study consists in setting an environment and the rules and observing the outcomes.

For several authors, these concepts justify the use of such experiments in economics. Indeed, experiments revealed the importance of institution and in the 1960s, from the results of the first experiments, economics transitioned from an Institution-Free theory to an Institution-Specific theory. Laboratory games showed that agents' behavior depends on how the institutional rules converts the actions (or messages) into outcomes. For example, the type of auction defines how the bids are converted into the price, and agents bid differently depending on this transformation (economists say they can be price-takers, price-makers or both).

Testing the impact of the institution is still one of the main purposes of conducting such experiments, as shown in recently published reports (Capra et al., 2005).

1.2 Purpose of experiments: test, failure and design

Nearly 50 years of participatory experiments in laboratories of economics combined with an inspiration from physics led key researchers to present extremely interesting remarks on why economists conduct experiments.

The first set of purposes is a physics-inspired approach. It consists in testing and comparing economic theories as well as comparing environments and institutions. This set is close to what multi-agent simulation experimenters call validation and consolidation of models.

Another purpose mentioned by Smith is to « explore the causes of a theory's failure ». Failure for computer scientists mostly means bugs and engineering problems, and when the simulation does not follow what was expected, we look for bugs or we modify the model. Economists have a deeper knowledge of discrepancies between what is predicted and what really happens in participatory experiments, and they teach us that the theory is not always faulty.

Reviews of experimental games useful for measuring social preferences by Camerer and Fehr (2003) show that in all cases, the predictions with rational and selfish players are not coherent with experimental results. Such a purpose is linked with the importance of the Duhem-Quine thesis within experimental economics methodological research. This thesis is often summarized as ``every experimental test of a theory is inevitably a joint test of theory conditional on the auxiliary hypotheses and an experiment's design and implementation". Authors consequently stress both the importance of the way in which experiments are conducted and the relativity of apparent failures of theories in experiments. Smith (1994) gives several examples of failures caused by hypotheses that were overlooked, such as the importance of the order of amount of money involved in the game.

Finally, the third set of purposes consists in testing institution design and public policies. Camerer and Talley (2004) compare this purpose with airplane wing designs in wind tunnels. This purpose might be the less positive purpose of experimental economics and can be linked with participatory prototyping in engineering.

1.3 Methodological hints

Several authors provide many methodological descriptions and hints (Friedman, 1994; Camerer, 2003; Camerer and Fehr, 2003) related to experiments in economics.

Participants often are undergraduate students who get paid for participating. One can currently read on the website of the Interdisciplinary Center for Economic Science at George Mason University:

> « If you are a George Mason University student and are interested in earning cash in an economic experiment, please sign up with our online recruiter. You will receive e-mail invitations to participate in experiments that fit your schedule. »

E-mail seems to have become the most widespread method to hire student participants. Economists compared the outcomes with students and workers, although studies with workers are more sparse, and they also noticed that students with a major in economics behave differently from students with another major. Some studies focused on atypical populations such as autistic participants (Hill and Sally, 2002). Even if economists do not turn into sociologists, they measure and acknowledge differences in the outcomes of experiments depending on the population involved in the game. Comparative studies of experiments are closer to anthropology and are presented with a strong anthopologist inspiration (Henrich et al., 2001).

To avoid a wide range of effects economists do not control, games are anonymous and the game is often played only once and usually excludes verbal communication between players. The games are also described with abstract symbols to avoid any effect of the vocabulary used. Studies showed that players behave differently depending on how things are called (Andreoni, 1995).

Camerer also notes that deceiving players is tabooed in experimental economics because players must believe the information given by the experimenters and the credibility of experimenters must not be affected in the long run. This taboo is strongly backed by economical theory and expressed with theoretical wordings.

2 Role-Playing Games and Multi-Agent Simulations (MAS/RPG)

Role-playing games and multi-agent simulations have recently been coupled in what is called the MAS/RPG methodology (Barreteau et al., 2001) or the GMABS methodology (Adamatti et al., 2005). This methodology is based on a parallel between the purposes of role-playing games and the purposes of multi-agent simulations, or more accurately, the purpose of this methodology directly inherits from the purposes of role-playing games in scientific and development processes.

2.1 The purposes of role-playing games

The use of role-playing games in scientific or development contexts is three-folded: training, research and policy making.

Training is the dominant use of role-playing games. In most cases, it is used with professionals in training sessions. Examples of such use actually also include policy exercises conducted to adapt to vulnerabilities (Toth, 1988). The methodology consists in writing scenarios including surprise-rich, but still plausible futures and to play them in an organizational setting reflecting the issues addressed. It can be simulations of evacuation or terrorist attack scenarios. Stakes are simulated.

Research use of role-playing games is what is closest to experimental economics. This use is based on the idea that during a game, witnesses can learn from players behaviors. The games are often set to test a theory or reply to a specific scientific question. In this type of game, the setting needs not to be realistic, however particular care is taken to know why participants came to play to avoid biases. While having fun can improve training efficiency, it is not, in this use, a goal of the game and it will be avoided unless it is used and analyzed as a sign. Stakes are often real, i.e. players can earn real money by participating, and such experiments are based on the idea that players come into the game with their own habits and strategy. Unlike experimental economics, role-playing games for research purpose are often ended with a debriefing session with participants.

Policy making role-playing games belong to the participatory action research tradition. In addition to individual learning, games can be used for social learning as it gathers players around a particular problem and it often requires discussion among them. Since favoring discussion is a key property of role-playing games, they can be used for discussion or for negotiation support (Tsuchiya, 1998). In that use, the stakes are fake but the setting is realistic. The setting is sometimes constructed by the players themselves as a first step to open the discussion. Another property of role-playing games often used in that case is that roles can be exchanged, and consequently favor negotiation by giving a better idea of other participants' point of view in real situations.

The three purposes are difficult to actually separate. Different actors of a role-playing game may stress outcomes of various categories depending on who they are. Experimenters may focus on the scientific research purpose of the game while participants would stress the importance of individual learning.

2.2 Combining role-playing games with multi-agent simulations

Role-playing games and multi-agent simulations are combined within the MAS/RPG methodology. Like most multi-agent simulation methodologies, this methodology is presented in an iterative way (figure 1).

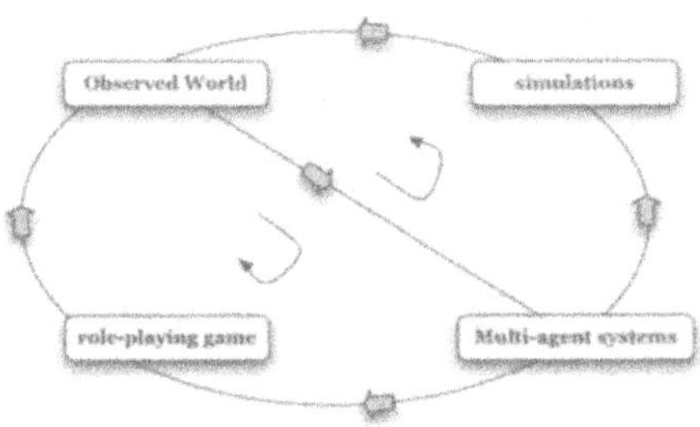

Figure 1: The MAS/RPG Methodology (from Barreteau et al. 2001)

The first step consists in developing a multi-agent simulation, usually to simulate the evolution of natural resources. Resources are often simulated by a cellular automaton model that may have been validated by domain experts. For example, MejanJeu (Etienne et al., 2003) is based on a validated model of pine trees invasion (Etienne, 2001). Stakeholders are represented by agents and the multi-agent simulation is designed to simulate various scenarios of resource management.

The second step consists in transforming the multi-agent simulation into a role-playing game where players will take the role of the agents within the simulation. The cellular automaton is often used during the game to simulate the evolution of natural resources. In most cases, the interface between the cellular automaton and the role-playing game is done by the experimenters.

The multi-agent simulation and especially the behavior of the actors is modified from what was learned by the experimenters during the role-playing game.

The purpose of this approach inherits from the purpose of role-playing games: well designed cellular automata and good scenarios can be used for training participants (Mathevet et al., 2005), the model of the multi-agent simulation is significantly improved from the outcomes of games played with stakeholders (Barreteau et al., 2001) and the system can be set to support negotiation (Etienne, 2003).

2.3 Comparison with experimental economics

The combination of role-playing games and multi-agent simulations can be compared with experimental economics. Experimental economics is tied as a discipline of economics to a very positive use, even if some experiments are close to participatory design of public policies and closely match the role-playing games used to test models. Moreover, experimental economics developed a finer description of what makes a model and its link with the theory that settles the use of participatory experiments to very specific tests.

The role of agents and of simulations of agents within the MAS/RPG methodology probably is underestimated. Most experiments following the MAS/RPG methodology can be described as either participatory design of multi-agent simulations, with outcomes concerning only the domain model of the simulation (to better understand stakeholders behaviors, for example) or as role-playing games where the setting is made more realistic and more suitable for participation by being backed with a tailored multi-agent simulation or cellular automaton.

Agents and multi-agent simulations can have more important roles in participatory simulations and this yields to what we call multi-agent participatory simulations.

3 Multi-agent participatory simulations

Multi-agent participatory simulations can be described as simulations where agents and players have similar roles. They can be seen as a direct extension of the MAS/RPG methodology. However, there are some major differences with role-playing games.

3.1 Methodology inspired from experimental economics

When used to extract knowledge from participatory experiments, the methodology of multi-agent participatory simulations is inspired from both multi-agent simulations and experimental economics. It is an iterative process with several loops (figure 2).

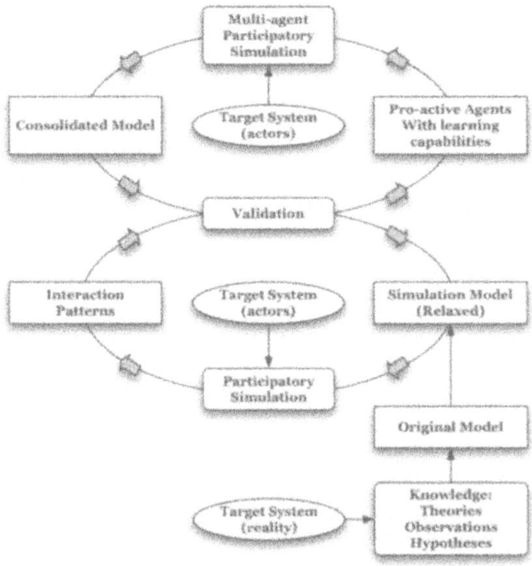

Figure 2: The Multi-Agent Participatory Simulation Methodology

The process begins with initial knowledge (theories) and an initial model. The model is then relaxed to be playable and modified for the purpose of the experiment. For example, to validate some hypotheses on the behavior of actors, the actions the actors can perform, what economists call the environment, must be broken into small primitives, such as analysis would reveal immediately if actors did not behave as expected. The institution, i.e. the rules of the game, is implemented into the role-playing game.

The first loop consists in running a participatory simulation which is very similar to role-playing games in the MAS/RPG terminology. The participatory simulation is mediated by computers and the experiments can be logged by log agents situated on the network and observing the game. These logs can then be processed and behaviors, in the form of interaction patterns, can be extracted. The outcomes of the simulation is used to validate the original hypotheses, as in experimental economics, together with the interaction patterns.

The second loop consists in building pro-active agents from the interaction patterns that were extracted during the first loop and to run multi-agent participatory simulations where the pro-active agents and the actors are playing together. The agents can learn from the behavior of human actors and the result can increase the understanding of the model, yielding to a consolidated model.

The first loop of this methodology was followed to conduct participatory experiments to study the coffee market in the state of Veracruz (Guyot and Drogoul, 2004).

3.2 Assistant agents

Designing simulations with agents and humans playing the same role that is complex enough to interest social scientists is a challenge. Indeed, without trying to fool humans into thinking that they play with other humans when they in fact interact with agents, we need to implement agents that are capable of participating to the experiments.

Our approach consists in designing assistant agents and to pair human players with such agents (figure 3).

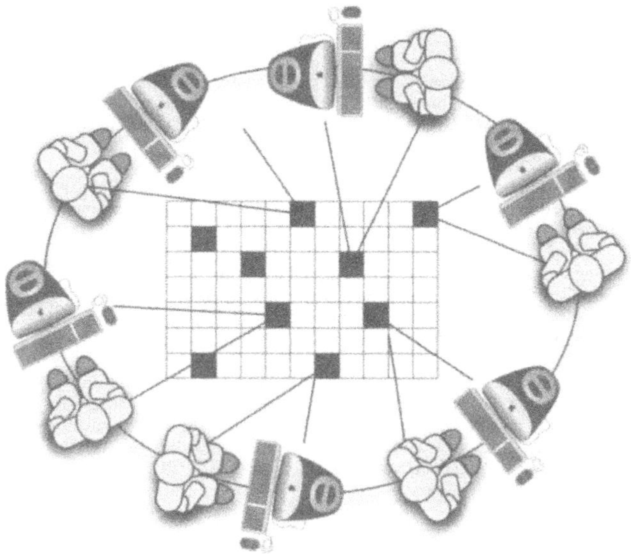

Figure 3: Players and assistant agents

Assistant agents need to have several capabilities. In particular, they need to be credible enough for human players to accept to play with them. To achieve this goal, we initialize assistant agents with the interaction patterns that were extracted from the first participatory experiments.

Their learning capabilities can also be used to better extract knowledge from the simulations. Their interaction with human actors includes asking questions to determine how the pair should behave within the simulation and asking the players to justify their choices. This technique actually consists in an artificial maieutic approach to extract players' knowledge (Sempé et al., 2005) and allows to more rapidly model the players behaviors than a pure observation in a role-playing game.

3.3 Participatory Design

Participatory design is one of the main goals of role-playing games and the motivation of few experiments of experimental economics. Likewise, multi-agent participatory experiments can be used for participatory design. This purpose can actually be achieved in two different ways.

The first method consists in organizing simulations to test new protocols or what economists call institutions. Nguyen-Duc actually conducted participatory experiments with assistant agents to test new protocols for air-traffic management (Nguyen-Duc et al., 2004). The advantages of such participatory simulations with assistant agents are two-fold: they collect immediate feedback from human users and one can set scenarios where agents suggest behaviors based on the new protocol to a human player or they can play simulations without any human user. Records of experiments following these scenarios can then be used during a debriefing session.

Multi-agent participatory simulations can also be used for participatory design of multi-agent systems. As in the MAS/RPG methodology, one can extract knowledge from stakeholders to build better multi-agent simulations. Besides, participatory experiments can benefit from involving humans in an emergence process and they can be used as a way to design socially inspired solutions to problems (Guyot and Drogoul, 2005). Recent experiments (Guyot et al., 2005) showed that multi-agent participatory simulations allowed to discover new behaviors and the multi-agent system obtained by implementing the strategies of human players had increased distributed solver capabilities.

Conclusion

Experimental economics has a long tradition of participatory experiments and provides us with many methodological and analytical hints on how to conduct such experiments. Their main purpose is mostly positivist: it consists in testing theories, even if economists qualify the importance of theory failures in participatory experiments by invoking the Duhem-Quine thesis.

Role-playing games and especially role-playing games coupled with multi-agent simulations is a first attempt to introduce agents within participatory experiments. Most experiments of this kind have a training or a participatory design purpose.

Even if both approaches are traditionally opposed in terms of scientific background, experimental economics stressing on the positive approach to test theories while role-playing games are justified by the need to analyze real complex phenomena, both provide interesting hints to build participatory simulations stressing on participatory design and on artificial maieutic.

We believe that making multi-agent simulation participatory not only reinforce the presence of agents within participatory simulations but also brings the development of participatory design of models and artificial maieutic to a further step. The technical reason for this improvement is that the loop is tighter: feedback from participants can be collected immediately through the interaction with assistant agents (instead of mainly using debriefing sessions in the MAS/RPG methodology or the outcomes of the experiment in economics).

As a consequence of the reinforcement of the presence of agents, multi-agent participatory simulations are closer to participatory design and experimental economics than role-playing games designed for training or policy-making. They can be situated between both approaches.

References

Adamatti, D.F., Sichman, J.S., Rabak, C., Bommel, P., Ducrot, R. and Camargo, M.E.S.A., 2005, Jogoman: A prototype using multi-agent-based simulation and role-playing games in water management. In: *Proceedings of the SMAGET-CABM-HEMA Conference.*

Andreoni, J., 1995, Warm-glow versus cold-prickle: the effects of positive and negative framing on cooperation in experiments. *Quaterly Journal of Economics* 110, 1–22

Barreteau, O. et al., 2003, Our companion modelling approach. *Journal of Artificial Societies and Social Simulation* 6

Barreteau, O., Bousquet, F. and Attonaty, J.M., 2001, Role-playing games for opening the black box of multi-agent systems: method and lessons of its application to senegal river valley irrigated systems. *Journal of Artificial Societies and Social Simulation* 4

Camerer, C.F. and Fehr, E., 2003, Measuring social norms and preferences using experimental games: A guide for social scientists. In Henrich, Boyd, Bowles, Camerer, Fehr, Gintis, McElreath, eds.: *Foundations of Human Sociality - Experimental and Ethnographic Evidence from 15 Small-Scale Societies.*

Camerer, C.F. and Talley, E., 2004, Experimental law and economics. In Polinsky, A., Shavell, S., eds.: *Handbook of Law and Economics.*

Camerer, C.F., 2003, *Behavioral Game Theory.* New York/Princeton University Press, Princeton, New Jersey

Capra, C.M., Tanaka, T., Camerer, C.F., Munyan, L., Sovero, V., Wang, L. and Noussair, C., 2005, The impact of simple institutions in experimental economies with poverty trap. Technical Report 0508, Department of Economics, Emory University (Atlanta)

Castro, B. and Weingarten, K., 1970, Towards experimental economics. *Journal of Political Economy* 78, 598–607

Etienne, M., 2001, Pine trees - invaders or forerunners in mediterranean-type ecosystems: a controversial point of view. *Journal of Mediterranean Ecology* 2, 221–232

Etienne, M., 2003, Sylvopast: a multiple target role-playing game to assess negotiation processes in sylvopastoral management planning. *Journal of Artificial Societies and Social Simulation* 6

Etienne, M., Le Page, C. and Cohen, M., 2003, A step-by-step approach to building land management scenarios based on multiple viewpoints on multi-agent system simulations. *Journal of Artificial Societies and Social Simulation* 6

Friedman, D. and Sunder, S., 1994, *Experimental Methods - A Primer For Economists.* Cambridge University Press, Cambridge

Guyot, P., Drogoul, A., 2004, Designing multi-agent based participatory simulations. In Coelho, H., Espinasse,

B., eds.: *Proceedings of 5th Workshop on Agent Based Simulations*, Erlangen, San Diego, SCS Publishing House, 32–37

Guyot, P., Drogoul, A., 2005, Two sides of emergence in participatory simulations. In: *Proceedings of Social Inspired Computing '05.*

Guyot, P., Drogoul, A., 2005, Using emergence in participatory simulations to design multi-agent systems. In: *International Joint Conference on Autonomous Agents and Multiagent Systems (AAMAS-05).*

Henrich, J., Boyd, R., Bowles, S., Camerer, C., Fehr, E., Herbert, G. and McElreath, R., 2001, In search of homo economicus: Behavioral experiments in 15 small-scale societies. *American Economic Review* 91, 73–78

Hill, E.L. and Sally, D., 2002, Dilemmas and bargains: Theory-of-mind, cooperation, and fairness. Working paper, University College London

Mathevet, R., Le Page, C., Etienne, M., Gigot, G., Lefebvre, G., Poulin, B. and Mauchamp, A., 2005, Butorstar: a role-playing game for collective awareness of reedbed wise use. Technical report, In prep.

Nguyen-Duc, M., Boucher, A., Drogoul, A., Duong, V., 2004, Towards participatory design of agent-based simulations an application in air traffic management. In Coelho, H., Espinasse, B., eds.: *Proceedings of 5th Workshop on Agent Based Simulations*, Erlangen, San Diego, SCS Publishing House

Sempé, F., Nguyen-Duc, M., Boucher, A., Drogoul, A., 2005, An artificial maieutic approach for eliciting experts' knowledge in multi-agent simulations. In: *International Joint Conference on Autonomous Agents and Multiagent Systems (AAMAS-05).*

Smith, V.L, 1994, Economics in the laboratory. *The Journal of Economic Perspectives* 8, 113–131

Toth, F.L., 1988, Policy exercises: objectives and design elements. *Simulation and games* 19, 235–255

Tsuchiya, S., 1998, Simulation/gaming as an essential enabler of organizational change. *Simulation and Gaming* 29, 400–408

Wadsworth, Y., 1998, What is participatory action research? *Action Research International* 2

Research on a Learning System toward Integration of Case Method and Business Gaming

Kenji NAKANO[1], **Shinako MATSUYAMA**[2], and **Takao TERANO**[3]
Dept. Computational Intelligence and Systems Sciences,
Tokyo Institute of Technology, Tokyo
4259 Nagatsuda-cho, Midori-ku, Yokohama, 226-8502 Japan
Tel/Fax:+81-45-924-5583
[1] kenji.nakano@samsung.com
[2] Shinaka.Matsuyama@jp.sony.com
[3] terano@dis.titech.ac.jp

Abstract. This paper reports on a development of business game, which aims at integrating Case Method and Business Gaming. This research proposes a new gaming structure model. In this model, we structuralize the decision-making area of corporate managers and build a framework to implement it as a business game. The model also contains a method to quantitatively express a corporate structure which is a mixture of middle-term business policies and short-term business operations. We use this model to develop a business game that simulates a case of "Asahi Super-Dry" to demonstrate that Business Gaming is able to deal with the qualitative decision-making area, as well as the conventional quantitative operation area. This paper also verifies the effectiveness of using Business Gaming in Case Study. In the experiments, undergraduate and graduate students who majors in business administration played the "Asahi Super-Dry Game", and we analyzed and evaluated the learning effects on this learning system.

1 Introduction

Gaming Simulation is an effective method to virtually experience and learn business schemes in the real world. It is recognized and used in a wide range of educational institutions [1]. Case Method of the Harvard Business School is also an effective educational method and is widely adopted in various fields in the name of Case Study [2]. These two education methods have in common and both belong to the "Experiential Learning" in the broad sense, which is a theoretical framework proposed by Kolb [3] et al. However, the directions of their basic approaches are totally reversing and have evolved separately making the most of their own advantages. There is no precedent report that attempts to integrate these two methods.

The subject of our research is to integrate these methods: Business Gaming and Case Study. We already had reported on the fruits of our research at KES2004 [4], which was about:

1) Construct a method to model the business actions of corporate managers in the areas of "decision-making" and "operations" separately in order to create a gaming scenario that is applicable to various cases,
2) Develop a versatile business simulator based on the model of "decision-making" area using GSSM(Graduate School of Business Sciences, University of Tsukuba)'s business game developing tools [5],

In this paper, we report on two additional research results on:

3) Developing a business game that simulate a case of "Asahi Super-Dry," and
4) Verifying the availability of the business game as a decision-making system based on the experimental games we conducted.

This paper consists of the following chapters: Chapter 2 provides overviews of our approaches 1) and 2). Chapter 3 explains about 3), the method for developing the business game that simulates "Asahi Super-Dry" case.

Chapter 4 describes the configuration of the learning system. Chapter 5 reports on the result and its verification of experimental games we conducted with the system. And then we conclude the discussion in Chapter 6.

2 Development of Business Simulator that Handles Cases

2.1 Challenges and Approaches

We have addressed to the challenges on integrating Gaming Simulation and Case Study as below:

1) These two methods deal with different types of information (Case Study mainly deals with qualitative information).
--> Convert the qualitative decision-making items into quantitative data,
2) Case Study needs to deal with the divergence between corporate manager's decisions and the executions through the organization, which often arises in actual corporate management. For example, the decisions made by a corporate manager of a large firm are represented as a broad operating policy. However, there is no guarantee that the organization conducts the operation in full accordance with the policy. We need to contrive the expressions of the game.
--> A corporate manager controls the organization indirectly with the operating policy, and the organization basically conforms to the policy but not necessarily in perfect coordination, and
3) Corporate managers are always required to attain the financial performance and are not valued with the results of corporate revolutions. We need to express both of the short-term profit and the long-term vision in the game.
--> We evaluate corporate managers with the financial performance in short-term and with the comprehensive measures, including the results of the corporate revolution, in long-term. The result of corporate revolution exerts a strong influence on long-term performance.

2.2 Framework of Simulator

The business simulator has a flexible framework to accommodate to diversified case scenarios. It has separated modules for the long-term corporate decisions (i.e. corporate revolution) and for the short-term operational decisions (i.e. investments for facilities or new product developments).

Figure 1 illustrates the framework of the decision-making module and the relations between the major functions.

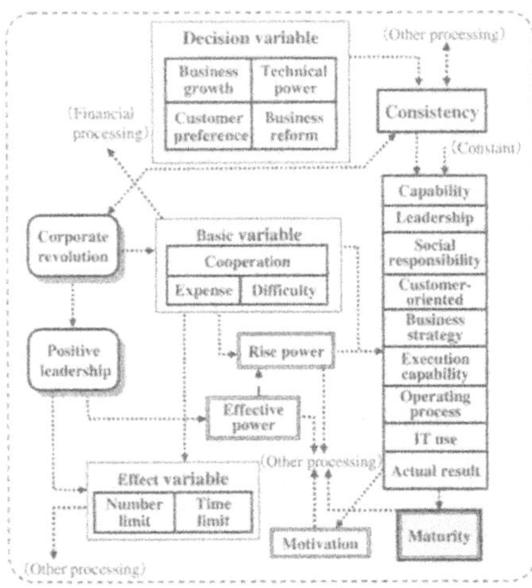

Fig. 1. Framework of decision-making module

3 Developing "Asahi Super-Dry" Game

3.1 Specification and Framework of the Game

The "Asahi Super-Dry" is a succeeded case of Asahi Breweries, Ltd. Asahi was a weak brand in Japanese beer market in 1980's, where Kirin Brewery Co., Ltd. retained its oligopolistic position. In the market, Asahi doubled its market share by developing a new taste beer "Super-Dry".

The business game that simulates this case (the "beer game") allows players (or learners) to make separate decisions "as the top management for business administrations" and "as the middle management for business operations". The players learn about the Asahi's succeeded case through the experiences in the quasi-market resulted from these decision-makings. Therefore, the requirements for the beer game are: it must be able to handle these two types of decision-makings, and it also needs to implement a model of a quasi-market where multiple companies with different management environment compete against each other.

The specification of the beer game need to incorporate mechanisms that:

1) reflect how much the top management's middle-term policy infiltrates into the operational tier in the lower organization, which indirectly exerts an influence on the implementation of the policy and the performance of the organization,
2) the performance of the organization reflects the consistency between the top management's policy and the middle-tier operations, and
3) differentiate the business environments and conditions of the competing companies at the beginning of and during the game (the existences of the other companies must be considered).

We have actualized the beer game using the models that:

1) adopt the "maturity" (from the framework of "Japan Quality Prize [6]") as an internal variable for reflecting the top management's policy and decision-makings, which impacts on the operations in middle-term,
2) provide separate entry screens for top and middle's respective decision-makings and check the "consistency" between them, which influences the operations in short-term, and
3) configure different initial values for every competitors so that each company starts the game from its unique condition.

3.2 Game Model and Decision-Making Items to be Learned

Figure 2 illustrates the game model of "Asahi Super-Dry." We take up the case of Asahi Super-Dry from "Strategic Organizational Innovation" (Kawai [7]) as our research subject. We have selected four of top management policies among the major management epochs of Asahi Breweries between 1982 and 1989: "share No.1", "customer-oriented", "R&D", and "corporate revolution". These are the decision-making items to be learned in the game.

We also determined five decision-making items for middle operations: "sales goal", "advertisement", "capital investment", "market research", and "new product development." In the gaming scenarios, the measured consistency between the top management policies and the middle operational decisions has an affect on the feasibilities of the operations.

Our game model distinctly separates the top management policies and the middle operational decisions, and therefore it forces players to make decisions in two different tiers. This model provides a style of learning decision-makings in "top" and "middle" tiers concurrently or individually. It also allows teams to learn from the differences in game results with or without good communication between "top" and "middle" within the teams. This novel style of the gaming enables players to extensively experience the way of making decisions in corporate managements.

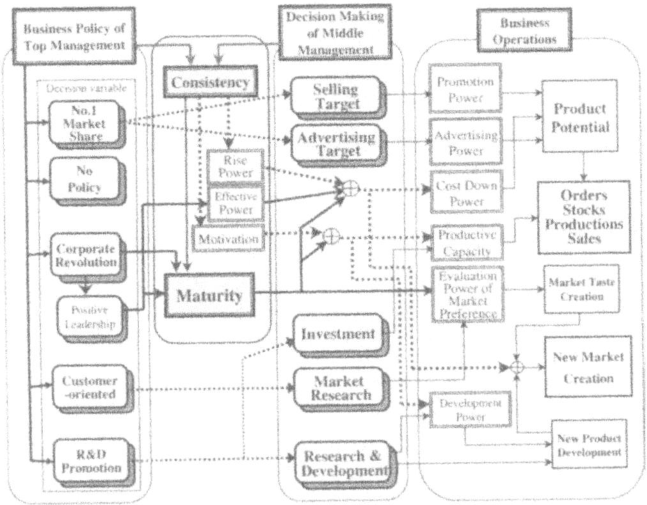

Fig. 2. The game model of "Asahi Super-Dry"

3.3 Conditions for Successful Decision-Makings

This beer game implements the succeeded decision-making processes of Asahi Super-Dry as the following success conditions:

- Process 1: <Corporate Revolution>

The corporate revolution aims at renovating the corporate culture in order to conquer the current market structure. The successful revolution leads to the successful development of the new taste beer, "Super-Dry", which creates a new market for the company. The time frame offered for the corporate revolution is two years. The revolution succeeds or fails depending on the top management's decision on "which capability of the organization should be improved and how far the leader should get involved with it."

- Process 2: <Market Research>

In order to make the market development successful, the players need to satisfy a certain levels of: "the consistency between the top management's policy and the middle-tier operations", "the maturity of the organization", "the capability for the customer relations", and "the cumulative investment for the market research."

- Process 3: <Research and Development>

To make the R&D successful, the players are required to fulfill a certain levels of: "the consistency between the top and middle decisions", "the maturity of the organization", "the capability of developing strategy", and "the cumulative investment for the R&D." Furthermore, the players need to continue the investment for more than two years, and they need to succeed in the market research to determine the customer preferences that are effective for R&D.

- Process 4-6: <New Product Development>, <Introduction to Market>, <Share No.1>

The new product development (the process 4) succeeds if the company succeeds in R&D and continues the R&D investment in the subsequent year. However, the development work will be suspended when a divergence arises between the top and middle decisions.

3.4 Gaming Design of Beer Game

Decision-making processes. Figure 3 illustrates the timing and the judgment criteria of six decision-making processes.

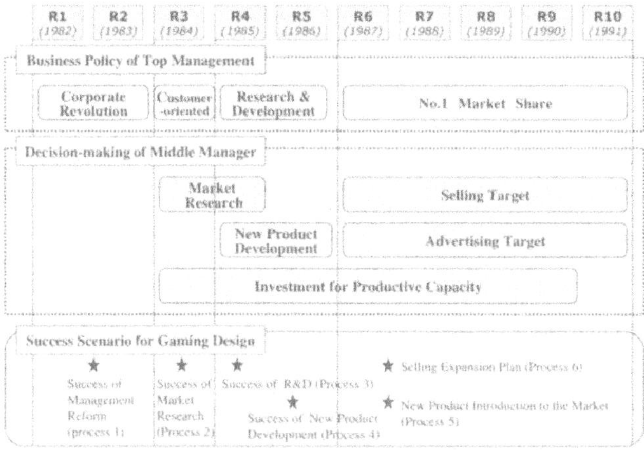

Fig. 3. The criteria of six decision-making processes

Market share. The market share of Asahi transits as:

- approximately 10% during 1982-1985,
- 15-16% in 1986, and
- 25-26% during 1987-1992.

The transition accommodates to the timings of the success of the new product development and its introduction to the market.

Verification of reproducibility. We check the reproducibility of the case of "Asahi Super-Dry" on the developed beer game by simulating a game using agent players. Player 1 (Asahi) is configured as the reference player who enters the correct values for the Asahi's succeeded processes. We configure the other players as Player 2-4 (Sapporo, Suntory, and Kirin) who use random input values that do not include the succeeded pattern.

As the result of the simulation, this evaluative play has reproduced the case of "Asahi Super-Dry." This result demonstrates the validity of the business game which we have developed based on the case.

4 Configuration of Learning System

We have developed a learning system based on the beer game, which allows learners to simulate the learning processes of Case Method. This system consists of two modules for individual and group learning. This learning system does not aim at simply replacing the well-established learning process of Case Method with business games. The purpose of the system is to explore the possibility of increasing the learning effect by appending business games to the learning process.

4.1 Development and Execution Environment of Learning System

We have developed the learning system using Business Model Description Language (BMDL) and Business Model Development System (BMDS) that are developed in GSSM, University of Tsukuba.
BMDL is a description language that enables us to represent own business models only with basic computer programming knowledge. A game developed with this language runs on BMDS that is constructed on a WWW server. BMDS stores the data resulted from a game (such as variable data) in the server as a tabular format file for each team. The group-learning version of the beer game is developed using BMDL and BMDS to allow multiple learners to concurrently learn decision-makings through the gaming style simulated experience.

4.2 Purpose and Configuration of Individual Learning Tool

The individual learning tool enables learners to understand and analyze the case individually before participating group learning. We have developed this tool based on BMDS and the agent technology [8]. A learner plays the game by making own entries (decision-makings), and the software agents plays the roles of the competitors with the pre-determined behaviors.
The learning objectives of this tool are to:

1) analyze the decision-makings from the succeeded (or failed) patterns of the case, and
2) plan the own decision-makings through the actual decision-makings in the games and the analyses of the results.

The original BMDS requires a facilitator who manages the process of the game. We have ameliorated the BMDS to allow a learner to proceed with the game without a facilitator so that the learner is able to study by oneself. We have pre-determined the other players' behaviors by describing the agents. We also have improved the user interface to better suit to self-learning by reforming the input and output screens and by adding more descriptive texts and the operating instructions.

5 Verification of Beer Game through Experimental Games

5.1 Purposes and Method of Experiment

The experimental games aim at verifying the following two points on learning decision-makings with the business game:

- the index and the methodology for quantitatively measuring the learning effects, and
- the effectiveness of the individual learning tool.

Sixteen learners have attended the experimental games. Four of them are adult students from the graduate school of Tsukuba University, and the rest of them are graduate or under-graduate students from the business administration department of Yokohama National University. They are also classified as:

- eight learners who had previously learned the game with the individual learning tool, and
- the other eight learners who have attended the group learning without any experience on the game.

We have divided them into four teams with four players each and played the game two times.

5.2 An Index for Quantitative Evaluation

To evaluate the result of the experiment, we adopt the concept of "area ratio" as an evaluation index. Figure 4 plots the results of two experimental games. The lines joining the plotted dots are called "tracking curves." The figure also contains the "reference curve" which represents the result when entering the correct value on every decision-making. We use the area ratio and the tracking curves in the following principles:

1) We define the "area ratio" as a ratio between the two areas: the area surround with the X-axis and a tracking curve of an experimental result, and the area surround with the X-axis and the reference curve,
2) Better curves moves upward, which indicates that the player reaches higher process levels, and
3) Better curves moves to the left, which indicates that the player reaches a certain process level earlier.

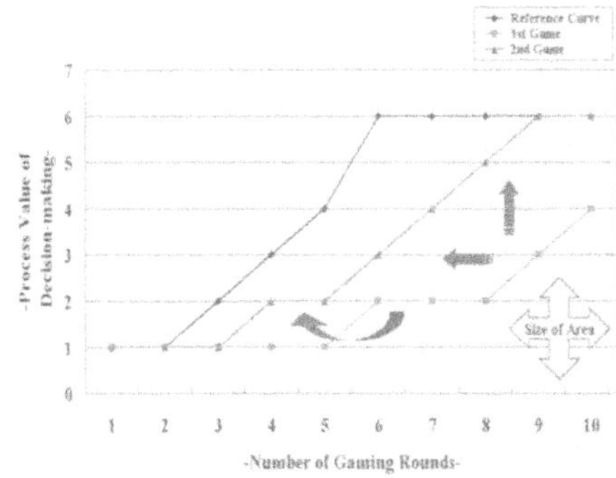

Fig. 4. Definition of area ratio

3.3 Result of Experiment

Figure 5 shows the number of the succeeded processes that each player has reached after ten rounds in the first and second games. The two horizontal lines in the figure indicate the average value of all players. Figure 6, the one similar to Figure 5, shows the area ratios and the averages in each game.

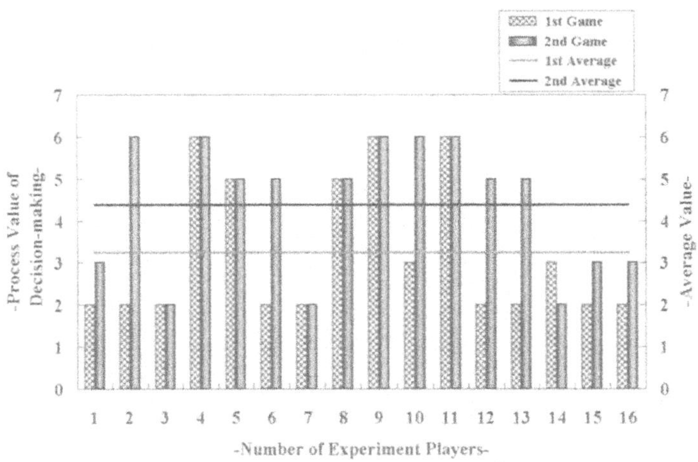

Fig. 5. Result of succeeded processes and the averages

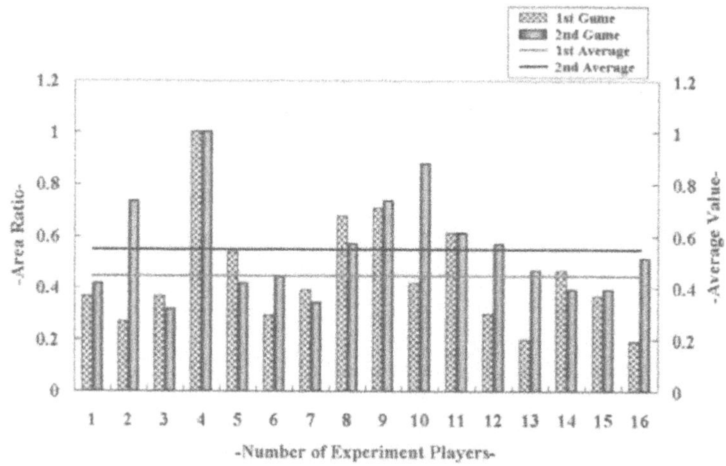

Fig. 6. Result of area ratios and the averages

The result shows that:

1) the average number of the reached processes has improved from 3.250 to 4.375, and
2) the average value of the area ratio has improved from 0.446 to 0.548,

in comparisons between the fist and second games. This indicates that the learning effect has become prominent by playing the game more than once.
We determine the learning effects in this experiment using the criteria below:

1) Larger number of reached processes indicates higher learning effects,
2) When multiple players have reached to the same process level, the player with larger area ratio is evaluated higher, and
3) Individual learners are evaluated based on the tracking curves of the decision-makings.

Table 1. Lists of the learning effects

Player Number	1st Game		
	Process	Maturity ↑	Area Ratio
4	6	592.96	1
9	6	483.84	0.707
11	6	508.28	0.61
8	5	548.45	0.676
5	5	673.36	0.537
14	3	479.8	0.463
10	3	561.6	0.415
7	2	211.28	0.39
1	2	457.97	0.366
3	2	328.29	0.366
15	2	408.6	0.366
12	2	399.02	0.297
6	2	540.95	0.293
2	2	494.02	0.268
13	2	431.93	0.195
16	2	350.94	0.189

Player Number	2nd Game			Improvement tendency	
	Process	Maturity ↑	Area Ratio		
4	6	583.63	1	→	
10	6	562.95	0.878	↑	
9	6	503.84	0.732	↑	
2	6	527.9	0.732	↑	
11	6	524.55	0.61	→	A
8	5	531.96	0.568	↓	
12	5	452.08	0.568	↑	
13	5	416.08	0.463	↑	
6	5	576.07	0.439	↑	
5	5	666.25	0.415	↓	
16	3	572.93	0.514	↑	
1	3	456	0.415	↑	B
15	3	510	0.39	↑	
14	2	203.68	0.39	↓	
7	2	148.55	0.341	↓	C
3	2	451.32	0.317	↓	

Table 1 contains the lists of the learning effects in two games that is organized based on the above criteria. These lists serialize the players by their evaluative scores of the learning effects (highly evaluated players are listed upwards). The right most column indicates the learning effect evaluation of each player between the first and second games. In this table, eight players (the Player 1 to 5, 9, 10, and 11) had previously learned the game with the individual learning tool.

5.4 Examination

Evaluation of Learning Effects. We examine the learning effects in the experimental games quantitatively, and we classify the players into the following three groups based on the Table 1 and the evaluation of their tracking curves.

<Rank A>

This group contains ten players who have reaped sufficient benefits of the leaning through the two experimental games.

<Rank B>

This group contains three players who have reaped limited benefits through the two games but will reap sufficient benefits if they continue the learning.

<Rank C>

This group contains three players who are in no situation to expect higher effects even if they continue this style of learning. We assume that they may have no aptitude for the learning of decision-making or for the gaming style.

Figure 7 shows the characteristic tracking curves that are selected as an example from each of three groups.

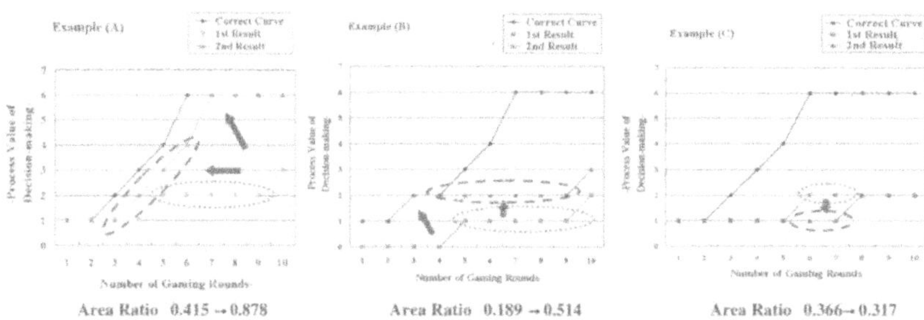

Fig. 7. Tracking curves for an example of three groups

The left chart shows the result of a player in <Rank A> group, who has improved the result remarkably on the second game as compared to the first game. The center chart shows the result of a player in <Rank B> group, who has reached only to the third process but has improved the value of the area ratio. The evaluation of his tracking curves also indicates significant improvement in the learning effect. The right chart shows an example from <Rank C> group, which indicates a reduced learning effect. In this example, the decision-making processes that the player has reached in the second game are lower or slower than the first game. The player has not been able to reproduce the result of the first game on the second game because the performance in the first game had been achieved without sufficient understanding of the game. We assume that the player was still in the process of learning by trial and error.

Effectiveness of Individual Learning Tool. In this section, we evaluate the effectiveness of preliminary learning on the individual learning tool based on the result of experimental games.

Table 2. The effectiveness of preliminary learning

Individual Learning Tool	Index	Average-value of 1st Results	Average-value of 2nd Results
Yes (8 Players)	Process Value	4	5
	Area Ratio	0.534	0.637
No (8 Players)	Process Value	2.5	3.75
	Area Ratio	0.359	0.459

Table 2 compares the averages of the quantitative values (described above in Table 1) between the eight players with the individual learning experience and the other eight players without it. As expected, the players with the individual learning experience have made better performance both on the number of reached processes and the value of area ratio.

6 Conclusion and Future Work

In this paper, we have proposed a new Business Gaming model, which enables generic and multilateral learning based on the simulated "case." This model is structured from three tiers that imitate the structure of business organizations: the top management decisions, the middle operational decisions, and the computer-automated implementations of the operations. We have developed a business game based on this model and a method that sets different initial values for each player at the beginning of games. The developed game has succeeded to reproduce the scenario of the successful "Asahi Super-Dry" case, which indicates the possibility for applying business games to Case Study. Then we have conducted experimental games and analyzed the result. One of the benefits of Business Gaming method is that it provides quantitative values as the results of games. We have evaluated its learning effects based on the comparisons of those quantitative values. The result of the analysis has demonstrated the usefulness of the evaluation method.

The result of the experiment reported in this paper suggests many challenges to be carried over for our future work. However, we have made significant progress by indicating the availability of integrating Case Study and Business Gaming.

References

1. Shirai,H.,et al.: Game Development Tool Kit for Business People in Japan. Simulation/Gaming, Vol.34, No.3, (2003) 437-446
2. Barnes, L.B., et al.: Teaching and the Case Method: Text, cases, and readings, 3rd edn. Boston, MA: Harvard Business School Press (1994)
3. Kolb,D.A.: Experiential Learning : Experience as the Source of Learning and Development. Englewood Cliffs. New Jersey. Prentice- Hall, Inc (1984)
4. Nakano,K., Terano,T.: From Gaming Simulation to Case Method -Empirical Study on Business Game Development and Evaluation-. KES2004 Lecture Notes in Computer Science, (*Proc.PartI*). Springer-Verlag, Berlin Heidelberg New York (2004) 472-479
5. Terano,T., et al.: Understanding your Business through Home-Made Simulator Development. Developments in Business Simulation and Experiential Learning, Vol. 26, (1999) 65-71
6. Japan Quality Award Council: Assessment Criteria of the Japan Quality Award 2003. Japan Productivity Center for Socio-Economic Development (JPC-SED) (2003)
7. Tadahiko,K.: Strategic Organizational Innovation. Yuhikaku Corp. Tokyo (1996) 35-57
8. Kobayashi,M., et al.: Application of Agent Technology in Business Game Development. The 16th Annual Conference of Japanese Society for Artificial Intelligence (2002) (3F2-01)

Market and Policy

Sequential bargaining on a perishable goods market: An agents-based Model.

S. Moulet and J. Rouchier

GREQAM, 2 rue de la Vieille Charité 13226 Marseille cedex 02

Introduction

Bilateral Sequential Bargaining has been extensively researched, for instance in game theory (Osborne and Rubinstein 1990, 1994, Muthoo 1999). The assumptions given in these models are quite far from the reality. In this paper we focus on an empirical market and try to find a model based on assumptions as close as possible to the reality. More precisely, we are interested in the (French) Fruit and Vegetable Market of National Interest in France.

In economics, different techniques are used to predict or understand the individuals' behavior on a market. Some analytical approaches study the behavior of sellers and buyers on a market considering that agents are perfectly rational but have access to limited information. The setting closest to our situation considers a seller and a buyer bargaining over the price of a good. The price that a seller is willing to accept is unknown to the buyer, and the price that a buyer is willing to pay is unknown to the seller at each point in time. The key assumption which is not valuable in our case is that the agents know the distribution from which the values are drawn (Myerson and Sattertwaite 1983).

Another technique that seems to be more suited to the study of the market we are looking at is the Multi-Agent Simulation (or Agent-Based Simulation). This tool enables us to study the outcome of the bargaining, in a context close to the context of the market we are interested in, considering heterogeneous agents characterized by evolving decentralized beliefs about their environment (Rouchier et al 1998, Rouchier et al. 2000). As expressed by Kirman and Vriend (2001): "To understand the dynamics of interactive market processes, and the emergent properties of the evolving markets structures and outcomes, it might pay to analyze explicitly how agents interact with each other, how information spreads through the market, and how adjustments in disequilibrium take place".

The market we are interested in is a face–to-face institutional form frequented by sellers (local producers or importer wholesalers) and buyers (retailers or caterers). Every day, the sellers display their stocks without disclosing the prices and wait for some potential buyers to bargain with. The exchanged information remains private. The individuals do not have certainty over the quantities and prices available on the market. They base their beliefs only on their own experiences and some public signals (average prices of sales during the previous days, weather conditions). Therefore, their reservation prices change according to individual experiences.

All these characteristics make the market very far from the basic setup assumptions in the models that we can find in the literature, and which are able to predict explicitly the outcome of negotiations. Our approach consists in, first, observing the individuals behavior on this market to build an agents-based model. Then, we conduct simulations to compare our model's results with some data and, more precisely, with some empirical stylized facts highlighted to validate our model (Kirman et al 2005). Lastly, we shed light on main characteristics which have an impact on the negotiation dynamics and knowledge acquisition. We are interested in the process that leads to transaction prices and, more generally to the establishment of a stable market. We want to find a behavioral process explaining the market state we observe.

Our work has been influenced by the following researches: the results of a survey conducted on the empirical market (Rouchier 2004; Kirman et al. 2005) and the representation of bargaining (Kirman and Vriend 2001; Brenner 2002; Rouchier 2004). Rouchier (2004) describes the same market as ours, but without reproducing any complex negotiations. Her paper gives us, first, a description of the real market which is a source of inspiration to build our model. Second, the fact that the model is based on the reality is very interesting and we adopted this approach in this paper. Brenner (2002) studies the dynamics of market prices under the assumption of behavioral learning by sellers and buyers. More precisely, he is interested in the question of whether bargaining of prices becomes common and whether the dynamics of prices converge. He simulates a learning process and compares his results to a game theoretic prediction under the assumption of rational agents.

In our paper, we find a learning process involving a convergence toward a stable state to which, in turn, we compare to the empirical facts. Our main results are consistent with the empirical market' dynamics: the sellers' negotiating power is always superior to the buyers and tends to decrease. The loyalty of buyers does not influence the transactions prices, and the frequency of each "nature of contact" is similar to the observed ones. Moreover, we observe that even if at the beginning, the sellers' initial beliefs are different they eventually converge. This remark is of extreme importance since we may now make the assumption that there exists homogeneity in the population of sellers on the market. This fact, in turn, gives more importance to the data base we use here. It seems to be acceptable to analyze only one seller since the homogeneity in the sellers' beliefs and experiences tends to be the same in the long-run.

The remainder of the paper is organized as follows. In section 2, we present the main stylized facts of the empirical market. In section 3, the agents-based model is described. Section 4 outlines the results of the agents based model. Lastly, section 5 concludes.

Stylized facts

The set of quantitative data we possess (already analysed by Kirman et al. (2005)) is a series of values of offers, counter-offers and transaction prices that were recorded on the market a few years ago. Even when no transaction took place, intermediate offers were recorded (giving series of offers and counter-offers), as was the absence of buyer (idle period). The series of prices (each one being referred to as a "sequence of negotiation") were obtained by observing one seller for 50 market days. Altogether, 3960 prices for 4 kinds of vegetables were gathered. This is a very original data set, since it shows intermediate offer prices, whereas usually only final prices are to be found in market data. These values can be of help when trying to analyse beliefs that actors have about the system at a given moment. In the paper we fix our attention on the data concerning the leeks. The leek is a product present on a market for only a short time. The buyers do not have certain information to evaluate its value. They try to get information from the sellers to evaluate as best as possible the conjuncture and so obtain the price as just as possible. With this data set it is possible to analyze the impact of information provided by the offers made during the sequences of negotiations. They reveal some insights on the evolution of the market, such as the fact that proposed prices and the index of sellers' bargaining power index decrease during the day (the bargaining power index is a parameter Kirman et al. construct to observe the relative deviation between the transaction price and the counteroffer with respect to the difference between the initial offer and the counteroffer). This data set and more precisely the statistic analysis made by Kirman and al. (2005) is used as a benchmark in our paper. Their paper is the main reference for our analysis. Thus, the reader is invited to refer to it.

Assumptions and Model

We consider a finite population of N_s sellers and N_b buyers with a bilateral relationship focused on the price of a single and perishable good available in limited quantity. We repeat 100 times the same scenario: Before the market opens, sellers buy their supply at an exogenous price which becomes their limit value denoted s_l. We denote r_i^0 to be the initial stock of seller i. Then, when the market opens, all buyers enter. Each of them needs one unit of good and chooses a seller according to his own experience and starts bargaining. The sellers run their stocks in the following way: Each day, they buy the quantity as sold the day before increased by a unit. This unit has a cost lower than the expected profit the buyer can have. Indeed, the matching process is random but takes into account the fact that the buyers visit the sellers with probabilities proportional to the discounted profit realized with the sellers in the past.

The choice of the sellers is based on the buyers' experiences with a discounting rate δ which describes the gradual forgetting of past events and ensures that information is more relevant for the current situation. A discrimination rate measures also the non linearity of the relationship between the probability to choose a seller and the profit realized with him. More precisely, the decision is based on the most common rule found in the literature: The Boltzmann Distribution Selection. We assume that, if several buyers visit the same seller, the seller handles them according to their indice ($\mathbf{1} \ldots \mathbf{N}_b$). Their bargaining follows a process to be described later on. At the end of one sequence of bargaining, if some buyers have not purchased the good, they can choose another seller until they have no time left to search. Otherwise they can leave the market.

Once matched, the seller makes a first offer. The buyer accepts or rejects it. If he rejects, he has the possibility to make a counter-offer and stop the bargaining process. If there is a counter offer, the seller can accept, reject and stop the bargaining process, or make a second offer that the buyer accepts or rejects. If no transaction happens, the agents separate.

The agents' procedures are inspired by the results of surveys carried out on the empirical market. We advice the reader to refer to Moulet and Rouchier (2005) for more details on the model. All buyers have 2 constraints. First, a time constraint which limits the number of opponents they can bargain with, and the length of the bargaining. We consider 2 costs: a search cost α and a bargaining cost β; Second, a budget constraint due to their limit values. The buyer knows the price at which he will sell the good to his own customers. All sellers have also a constraint value: They cannot sell at a price lower than their own supply price denoted b_l. This rule is directly due to the engagement taken by the sellers who certify to dispose of the goods instead of selling them at a price lower than the purchase price.

The agents have some beliefs (which evolve over time) on the willingness to pay or to accept an offer. We denote by s_i^t the beliefs in the period t of seller i. The buyer's beliefs are represented by a uniform distribution defined on a support which evolves over time. When a buyer shows up, the seller i offers the price that he thinks that the buyer can accept (s_i^t). The buyer accepts if the price is below his current reservation price (average beliefs). Otherwise, he decides if he wants to bargain with the seller or to leave. He does not make a counteroffer if his expected profit is bigger if he stops the bargain and visits other sellers than if he continues to bargain with him. The calculation of the expected profits is based on the idea that by staying with the same seller, the buyer can obtain the average between the offer and the lower limit of his beliefs. If he stops the bargaining, the future offer depends on the relative time he can stay on the market and the offer made by the seller. The seller accepts the counter offer if its value is higher than 80% of his reservation price. Lastly, the seller decides if he wants to makes the second offer. Even if there is no cost to the bargain, he decide not to make a second offer if the buyer's counter offer is too far from his beliefs (more than 20% below his reservation price). In this way, he sends the signal that the offer his too low. Otherwise, the second offer of the seller is lies between the first and the counter offer. Its exact value depends positively on the ratio of the remaining time and the remaining stock of the seller.

After each offer, the agents update their beliefs which represent the current situation. The buyers react directly after a signal. They update their beliefs by truncating the support of their beliefs or decreasing the lower limit. They are ready to make fewer concessions for the second offer at the end of the day than at the beginning of the day. The sellers update their beliefs at the end of each negotiation only when they observe a repeated behaviour. If more than 5 buyers act in the same way, they increase (resp. decrease) their beliefs. More precisely, after 5 [1]acceptances (resp. rejects) (after the first offer, after the counteroffer and after the second offer) they increase (resp decrease) their beliefs by a fixed amount. (10, 5 and 2%)

Results and calibration of the model

We study the relevance of the learning process given in the previous section by testing if the simulated market exhibits, in the long run, the same characteristics as the MIN in Marseille. More precisely, we are interested in the dynamics of the offers made during the negotiations, the transaction prices that arise in the case of transactions, and the variation of the length of the bargaining over the day.

We use the results of simulations for the last 20 periods and the same indicator as Kirman et al. (2005) in order to compare the data. Given the huge number of parameters, we fix some parameters and test the sensitivity of the model to alternatives as in the following three cases. When we know the empirical value of a parameter, we give it to the parameter (Ns and Nb, r_i^0). When we do not have the values but can find in the earlier research, we take standard values commonly used in the literature to match usual choices (α, β, δ). Lastly, concerning such parameters as the initial reservation value or the time constraint, we study their effects by scanning the plausible set of parameters. Given the lack of space, we indicate only the values that give the lead to the most consistent results. Hence, we fix Nb=1500 Ns=50, r_i^0 €[30,33], α=0.25 and β=0.1 and, without other specifications the initial limit values are 4 for the sellers and 5 for the buyers.

With the "homogeneous values simulations" we note that the initial sellers' reservation values have no impact in the long-run and, hence, neither has the difference of knowledge between buyers and sellers. All simulations produce results similar to the empirical data concerning: the sellers' bargaining power index (fig. 1) and the frequency of the natures of meetings (fig. 2). Indeed, we can observe in our simulations the decreasing variation of the sellers' bargaining power index over the trading day and the superiority of the sellers' bargaining power index compared to the buyers' one. The frequencies of contact with and without negotia-

[1] The value 5 has not an impact on the simulated data but only on the speed of convergence.

tion are very similar to the data set which corresponds to the interactions in the leeks market (Kirman et al.(2005)) : 60% of contacts led to a transaction in the simulated and the empirical situations. Concerning the first visited seller, we reproduce the same frequency of transactions without bargaining for the interactions in the leek market (around 20%). We reproduce the decrease frequency of transactions without bargaining, and the increase of the frequency of transactions with negotiation between the first and the second visited seller. Nevertheless, we do not reproduce the same frequencies for the second part of the day. This comes from the fact that we under evaluate the value of the disagreement outcome. The buyer does not give enough value to the good at the end of the day.

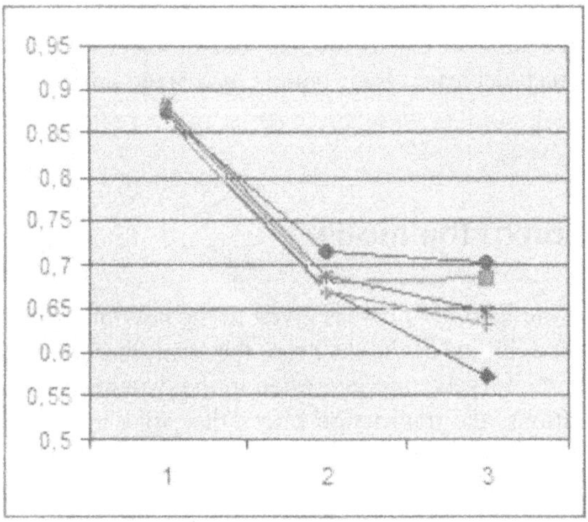

Fig.1 . Variation of the sellers' negotiation power for all simulations: Different plots correspond to different simulations. X-axis represents the parts of the day.

Table 1: Relative frequency of different nature of encounters for the simulated and the empirical data

	OO	OC	TNB	TWB
All simulated day	24	19	15	42
All empirical day	19.73	10.12	20.75	49.40
First simulated part of the day	31	7.81	21	40.19
First empirical part of the day	20.76	10.38	21.61	47.25
Second simulated part of the day	3	40	5	52
Second empirical part of the day	16.35	8.65	16.35	58.65

OO: only offer without transaction, OC: counter offer but without transaction, TNB: transaction but no bargaining and TWB: transaction with bargaining

In the "heterogeneous values simulation" the initial reservation values of sellers do not affect the long run aggregate state of the market. The simulation displays common features with the empirical data: 1) the ***succession patterns of the different natures of meetings*** are reproduced; 2) the ***change between the first and the***

second market of the day is clearly emphasized; 3) the ***decrease in the number of contacts with and without negotiation*** is reproduced. Nevertheless, we observe that the variation of the sellers' bargaining index is not strictly decreasing but always higher than that of the buyers.

These two sets of simulations show that, in the long run, systems do not depend on the sellers' initial beliefs. The learning process gives the agents the ability to adapt their beliefs to changes in the environment. This conclusion gives more importance to the data set analyzed in this paper: beforehand, we could have through that the seller, whose data are used, is different from other sellers in experience, and, hence belief. Our model, showing the convergence of beliefs regardless the initial reservation value, could indicate that our seller's data can be considered as representative.

The third set of simulations shows the influence of the time buyers can spend on the market. The simulation which results are nearest to the empirical data is the one where the sellers stay longest on the market: the frequency of the different nature of contacts is well reproduced; the bargaining power index decreases. The value to a buyer for 4 visits to sellers each day is in line with the qualitative survey conducted on the market. Hence, given the values of parameters, the learning process gives results close to the empirical observations.

Conclusion

Our model was built to represent interactions that take place on a specific market, about which qualitative and quantitative data had previously been gathered. The market is characterized by decentralized sequential bargaining through which individuals learn about the prices and quantities available on the market and adapt their behavior as the market evolves. Our interest is unusual, since we want to focus on the nature of interactions and the evolution of the bargaining power, the elements that are as important as the evolution of prices. In that sense, the model reproduces rather well the empirical data, given parameters. The general result is that, in any case, initial values of beliefs and reservation values have little, if no, influence on prices and or the loyalty of buyers. This first stage of calibration and test of sensitivity to parameters was hence, fundamental.

In the future, we wish to study the learning algorithm and the importance of the global structure of the market. For the moment, the global structure (here the ratio of the number of sellers and the number of buyers) was fixed according to the empirical market. This algorithm could also be tested technically, with variations of the speed of learning of sellers. Hence, the next step will be to better understand this system, hoping that it can be adapted to the representations of markets that are different from the one upon which we have initially based it.

Aknowledgements

We wish to thank Alan Kirman for his thoughtful remarks and, more generally, colleagues from the GREQAM who helped to structure this research. Anonymous reviewers at WEHIA and AESCS conference in 2005 made extremely useful comments on previous versions of this paper, and organisers of the Porquerolles summer school on the Agents-Based Models provided a rich framework for considering our research.

Eventually, Juliette Rouchier wishes to thank Toru Ishida and JSPS for allowing her to undergo her current research and Sonia Moulet wishes to thank the French Ministry of Research for providing financial help.

References

Aghion P, Bolton P, Harris C and Jullien B (1991) Optimal Learning by Experimentation, Review of Economic Studies 58(4): 621-654.

Brenner T (2002) A Behavioural Learning Approach to the Dynamics of Prices. Computational Economics 19:67-94.

Kirman A., Vriend N (2001) Evolving Market Structure: An ACE Model of Price Dispersion and Loyalty. Journal of Economic Dynamics and Control, pp. 459 – 502.

Kirman A, Schulz R, Hardle W, Werwatz A(2005) Transactions that did not happen and their influence on prices. In Press, Journal of Economic Behaviour and Organisation

Moulet S, Rouchier J (submitted) The influence of sellers' beliefs and time constraint on a sequential bargaining in an artificial perishable goods market. Journal of Economic Dynamics and Control

Muthoo A (1999) Bargaining Theory with applications Cambridge and New York: Cambridge University Press.

Myerson and Sattertwaite (1983) Efficient Mechanisms for bilateral trading. Journal of economic Theory 265-281

Osborne M. J, Rubinstein A (1990) Bargaining and Markets. Academic Press, San Diego, CA.

Osborne M. J, Rubinstein A (1994) Bargaining and Markets. The MIT Press, Cambridge,MA.

Rouchier J. (2004), Interactions routines and selfish behaviours in an artificial market, WEHIA 2004, Kyoto.

Rouchier J. and Mazaud J.P (2004) Trade relation and moral link in transactions among retailers and wholesale sellers on the Arnavaux market. Conference Proceedings of the Association for Social Economics, Albertville, June 2004.

Rouchier J., Barreteau O., Bousquet F (1998). Evolution and coevolution of individuals and groups. In: Demazeau (eds) Proceedings of the Third International Conference on Multi-Agent Systems, IEEE, Los Alamitos, pp. 254-260

Rouchier J, Bousquet F, Barreteau O., Le Page, C., Bonnefoy, J.-L. (2000). Multi-Agent modelling and renewable resources issues: the relevance of shared representations for interacting agents. In: Scott Moss (eds), Proceedings of MABS, Springer -Verlag, pp 181-197

Segal IR (2003) Optimal Pricing Mechanisms with Unknown Demand, American Economic Review 93(3):509-529

The Relationship between Relative Risk Aversion and Survivability

Shu-Heng Chen[1] and Ya-Chi Huang[2]

[1] AI-ECON Research Center
Department of Economics
National Chengchi University
Taipei, Taiwan
chchen@nccu.edu.tw

[2] Department of International Trade
Lunghwa University of Science and Technology
Taoyuan, Taiwan
yachi@aiecon.org

Summary. As a follow-up to the work of [4] and [5], this paper continues to explore the relationship between wealth share dynamics and risk preferences in the context of an agent-based multi-asset artificial stock market. We simulate a multi-asset agent-based artificial stock market composed of heterogeneous agents with different degrees of relative risk aversion (RRA). A wide range of RRA coefficients has been found in the literature, and so far no unanimous conclusion has been reached. The agent-based computational approach as demonstrated in this paper proposes the possibility that in reality there may be such a wide *survival range* of the RRA coefficient. In addition, the time series plot of the wealth share dynamics indicates that the higher the risk aversion coefficient, the higher the wealth share. This result combined with our earlier result ([5]) well articulates the contribution of risk aversion to survivability.

Keyword: Risk Preferences, CRRA (Constant Relative Risk Aversion), Blume-Easley Theorem, Agent-Based Artificial Stock Markets, Genetic Algorithms

1 Motivation and Introduction

The contribution of *risk preference* to the survivability (wealth share) of investors has recently received a series of theoretic and simulation studies (e.g., [1], [19], [18], [2], [4], [5]). The results are mixed, depending on how we approach this issue. While the standard analytic approach proves the irrelevance of risk preference to survivability ([18], [2]), the agent-based computational approach indicates the opposite ([4], [5]). This kind of inconsistency, as quite often seen in the agent-based computational economics literature, simply reflects the sensitivity of the classical (analytical) results to the interacting heterogeneous boundedly-rational behavior.

[4] actually supports an earlier result obtained in [1], which is also known as the *Kelly criterion* in financial economics. This result basically points out the optimal type of risk preference, namely, the *CRRA (constant relative risk aversion coefficient) agent with an RRA coefficient of one.* Equivalently, it is the log utility function. [5] reestablishes this result, while in an agent-based computational setting. They examine the long-run wealth share dynamics of eleven different types of CRRA agents, with RRA coefficients ranging from 0 to 1 with increments of 0.1. They find that in finite time the wealth share is positively related to the CRRA coefficient, and in the long run, only the agents with high CRRA coefficients can survive. All others become extinct.

This paper is an extension of [5] in the sense that we wish to extend the earlier testing domain of the CRRA coefficient from [0, 1] to an even larger positive domain. In doing so, we are inquiring whether a higher degree of risk aversion can actually enhance the survivability of agents. Notice that the degree of risk aversion is not the original concern of either the Kelly criterion or the Blume-Easley theorem ([1]), both of which are only concerned with the dominance of the log-utility type agents. Risk aversion is involved because the log-utility agent is also known as a CRRA type of agent with an RRA coefficient of one. Now, is this the optimal degree of risk aversion? Will more risk-averse agents (i.e. those with RRA coefficients greater than one) be also driven out of the market when they are competing with the log-utility agents? Or, would higher risk aversion help them survive? These are the questions that we try to answer in this paper.

We consider these questions to be particularly relevant because the empirical literature actually suggests a large range of relative risk aversion coefficients. Some of them are exactly one or less than one, but many more are greater than one. Of course, it is doubtful whether one can directly compare our results with those empirical values, since they refer to quite different stories. However, given the prevailing empirical results on high risk aversion, it is definitely useful to know what makes them so, and our agent-based computational setting can serve as a good starting point.

The remainder of this paper is organized as follows. Section 2 gives a brief introduction to a simple multi-asset model, which is originally used in [1] and later extended and modified by [18]. Section 3 presents the artificial multi-asset market, which is an agent-based version of the analytical model presented in Section 2. Section 4 gives the experimental design. To justify the range of the relative risk aversion coefficient considered in this paper, it starts with a brief review of the literature on the empirical estimation of the RRA coefficient in Section 4.1, followed by the setting of other control parameters in Section 4.2. The simulation results are provided in Section 5, and are followed by the concluding remarks in Section 6.

2 A Simple Multi-Asset Model

The simulations presented in this paper are based on an agent-based version of the multi-asset market as [1] and [18] have studied. The market is complete in the sense that the number of states is equal to the number of assets, say M. At each date t, the outstanding volume of each asset is exogenously fixed at one unit. There are I investors in the market, each indexed by i. At time t asset m will pay dividends w_m if the corresponding state m occurs, and 0 otherwise. The behavior of states follows a finite-state stochastic process, which does not have to be stationary. The dividends w_m will be distributed among the I investors proportionately according to their owned share of the respective asset. The dividends can only be either re-invested or consumed. Hoarding is prohibited. If agent i chooses to consume c, her satisfaction is measured by her utility function $u(c)$. This simple multi-asset market clearly defines an optimization problem for each individual.

$$\max_{\{\{\delta^i_{t+r}\}^{\infty}_{r=0},\{\alpha^i_{t+r}\}^{\infty}_{r=0}\}} E\{\sum_{r=0}^{\infty}(\beta^i)^r u^i(c^i_{t+r}) \mid B^i_t\} \tag{1}$$

subject to

$$c^i_{t+r} + \sum_{m=1}^{M} \alpha^{i,*}_{m,t+r} \cdot \delta^{i,*}_{t+r} \cdot W^i_{t+r-1} \leq W^i_{t+r-1}, \quad \forall r \geq 0, \tag{2}$$

$$\sum_{m=1}^{M} \alpha^i_{m,t+r} = 1, \quad \alpha^i_{m,t+r} \geq 0, \quad \forall r \geq 0. \tag{3}$$

In Equation (1), u^i is agent i's temporal utility function, and β^i, also called the discount factor, reveals agent i's time preference. The expectation $E(\ )$ is taken with respect to the most recent belief B^i_t, which is a probabilistic model used to represent agent i's subjective belief regarding the stochastic nature of the state. The maximization problem asks for two sequences of decisions, one on saving, and one on portfolios, as denoted by

$$\{\{\delta^i_{t+r}\}^{\infty}_{r=0}, \{\alpha^i_{t+r}\}^{\infty}_{r=0}\},$$

where δ^i_t is the saving rate at time t, and

$$\alpha^i_t = (\alpha^i_{1,t}, \alpha^i_{2,t}, ..., \alpha^i_{M,t})$$

is the portfolio comprising the M assets.

Equations (2) and (3) are the budget constraints. W^i_t is the wealth of agent i at time t, which is earned from the dividends paid at time t. Notice that these budget constraints do not allow agents to consume or invest by borrowing.

The equilibrium price $\rho_{m,t}$ is determined by equating the demand for asset m with the supply of asset m, i.e.

$$\sum_{i=1}^{I} \frac{\alpha_{m,t}^{i,*} \cdot \delta_t^{i,*} \cdot W_{t-1}^i}{p_{m,t}} = 1, \quad m = 1, 2, ..., M. \tag{4}$$

Rearranging Equation (4), we obtain the market equilibrium price of asset m:

$$p_{m,t} = \sum_{i=1}^{I} \alpha_{m,t}^{i,*} \cdot \delta_t^{i,*} \cdot W_{t-1}^i. \tag{5}$$

3 The Agent-Based Multi-Asset Artificial Stock Market

An *agent-based computational version* of the Blume-Easley-Sandroni standard multi-asset model is developed in [4]. One of the mainstays of the agent-based computational economics is *autonomous agents.* ([20]) The idea of autonomous agents was initially presented in [12]. Briefly, these agents are able to learn and to adapt to the changing environment without too much external intervention, say, from the model designer. Their behavior is very much endogenously determined by the environment which they are interacting with. Accordingly, it can sometimes be very difficult to trace and to predict the resulting outcome, known as the *emergent behavior.*

In this paper, we follow what was initiated in [12], and equip our agents with the genetic algorithm to cope with the finite-horizon stochastic dynamic optimization problem, (1) – (3). The GA is applied here at two different levels, a high level (learning level) and a low level (optimization level). First, at the high level, it is applied as a *belief-updating scheme.* This is about the B_t^i appearing in (1). Agents start with some initial beliefs regarding state uncertainty which are technically characterized by parametric models, say, Markov processes. However, agents do not necessarily confine themselves with just stationary Markov processes. Actually, they can never be sure whether the underlying process will remain unchanged over time. So, they stay alert to that possibility, and keep on trying different Markov processes with different time frames (time horizons). Specifically, each belief can be described as "a kth order Markov process appearing over the last d days and may continue". These two parameters can be represented by a binary string, and a canonical GA is applied to evolve a population of these two parameters with a set of standard genetic operators.[3]

Once the belief is determined, the low-level GA is applied to solve the stochastic dynamic optimization problem defined in (1) – (3). Basically, we use Monte Carlo simulation to generate many possible ensembles consistent with the given belief and use them to evaluate a population of investment plans composed of a saving rate and a portfolio. GA is then applied to evolve this population of candidates.[4]

[3] Details can be found in [4], Appendix A.2.

[4] Details can be found in [4], Appendix A.1.

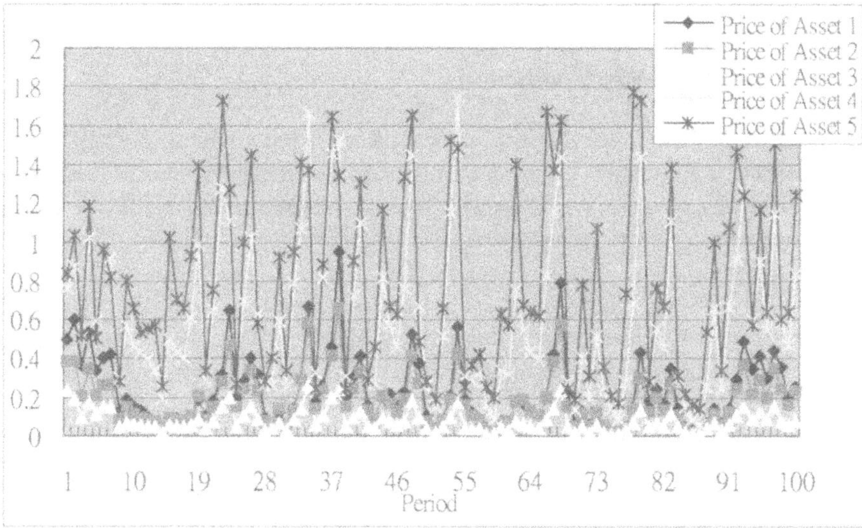

Fig. 1. Time Series Plot of the Prices of Assets; M =5.

In sum, the high-level GA finds an appropriate belief, and under that belief the low-level GA searches for the best decisions regarding savings and portfolios. This style of adaptive design combines *learning how to forecast* with *learning how to optimize*, a distinction made in [3]. These two levels of GA, however, do not repeat themselves with the same frequency. As a matter of fact, the belief-updating scheme is somewhat slow, whereas the numerical optimization scheme is more frequent. Intuitively, changing our belief in the meta-level of the world tends to be slower and less frequent than just fine-tuning or updating some parameters associated with a given structure. In this sense, the idea of *incremental learning* is also applied to our design of autonomous agents.

To simulate this agent-based multi-asset artificial stock market, a software called *AIE-ASM Version 5.0* is written using *Delphi, Version 7.0.* In each single run, we generate a series of artificial data. At the micro level, it includes the dynamics of agents' beliefs, investment behavior, and the associated wealth

$$\{B_t^{i,*}, \delta_t^i, \alpha_t^i, W_t^i\}_{t=1}^{100}, \quad i = 1, ..., I.$$

At the aggregate level, we observe the asset price dynamics

$$\{\rho_{m,t}\}_{t=1}^{100}, \quad m = 1, ..., M.$$

Figure 1 displays the time series plot of prices in a five-asset market. In this specific simulation, the state follows an *i.i.d.* process.

4 Experimental Design

4.1 Empirical RRA Coefficients

This paper and the coming experimental design are both very much motivated by the existing intensive empirical studies on the RRA coefficient. Therefore, we would like to give a brief survey here to present a general flavor on the wide dispersion of various estimates. For convenience we shall denote the RRA coefficient by ϱ.

Let us look at this from the left extreme, i.e. $\varrho \leq 1$. While [5] already questions the survivability of the CRRA agents with ϱ being less than 1, empirical studies supporting small ϱ still exist. The lowest value of ϱ is found in [16], which is only 0.3. A little above this is [11]'s estimate, that ranges between 0.3502 and 0.9903. The one that seems to best fit our early simulation result of [5] is [8], who find values of RRA of around unity. Instead of using real economic data, [10] derived their estimate from a laboratory experiment with human subjects, which lay between 0.6 and 1.4. At the other extreme, there are cases of $\varrho \geq 10$, such as 18 in [17] and 30 in [14]. These estimates are so large and are puzzling to many economists, having become part of the very famous *equity premium puzzle.* Next to those big ones, [13] also find a double-digit RRA in their estimates, which is close to 12. Despite these two extremes, most estimates suggest a moderate range for ϱ, and include the [1,3] interval ([9]), the [2,3] interval ([15]), [21]), the [2,5] interval ([6]), etc.

4.2 Control Parameters

According to the brief review above, we have set ϱ to lie in the [0.5, 5] interval. It starts from 0.5 with increments of 0.5 and continues up to 5. In this setting, we have a total of 10 types of CRRA agents, and assign 5 market participants to each type of agent. Hence, there are fifty agents in the market in total. There are also 5 assets available in the market ($M = 5$), corresponding to 5 states. Asset m pays dividends $6 - m$ ($m = 1, 2, ...5$). Two stochastic processes are considered in the experiments, namely, *iid* and the *first-order Markov.* Each is employed for one half of the total number of runs. The parameters of these two stochastic processes are also randomly generated in such a way that the axioms of the probability function are satisfied. Parameter values for the low-level and high-level GA can be found in [4], 4.1. We have 100 independent runs of the same experiment, and each run lasts for 100 market periods (T=100).

5 Simulation Results

Figure 2 shows the time series of the wealth share among the 10 types of agents. Notice that each line is based on an average of 100 simulations. The diagram seems to clearly indicate that the higher the RRA coefficient, the

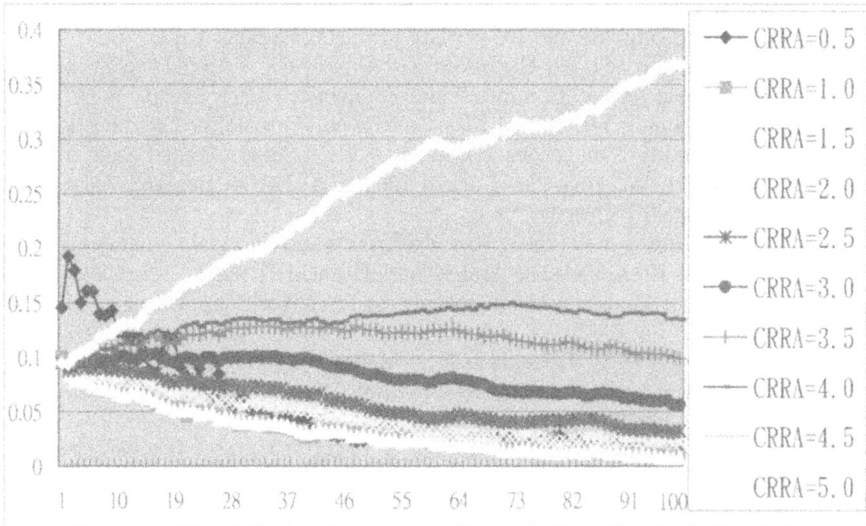

Fig. 2. Wealth Share Dynamics

higher the wealth share. For example, the wealth share of the agents with the highest risk aversion coefficient ($\varrho = 5$) shows strong growth from 10% in the initial period to 35% in the final period. Unlike what the Blume-Easley theorem ([1]) predicts, the log-utility agents ($\varrho = 1$) do not survive well in this case. Instead, their wealth share keeps on declining toward zero. Agents with lower values of ϱ share the same destiny. They are all dominated by agents with high risk aversion coefficient.

6 Concluding Remarks

This study can be possibly related to the extensive literature on the empirical estimation of the RRA coefficients. A wide range of RRA coefficients has been found in the literature, and so far no unanimous conclusion has been reached. The agent-based computational approach as demonstrated in this paper proposes the possibility that in reality there may be such a wide *survival range* of the relative risk aversion coefficient. Furthermore, the agents' wealth share is positively related to their RRA coefficients. This result combined with our earlier result (Figure 3, [5]) well articulates the contribution of risk aversion to survivability.

References

1. Blume L, Easley D (1992) Evolution and Market Behavior. Journal of Economic Theory 58: 9–40.

2. Blume L, Easley D (2001) If You're So Smart, Why Aren't You Rich? Belief Selection in Complete and Incomplete Markets. Working paper.
3. Bullard J, Duffy J (1999) Using Genetic Algorithms to Model the Evolution of Heterogeneous Beliefs. Computational Economics 13 (1): 41–60.
4. Chen S-H, Huang Y-C (2004) Risk Preference, Forecasting Accuracy and Survival Dynamics: Simulations Based on a Multi-Asset Agent-Based Artificial Stock Market. Working Paper Series 2004-1, AI-ECON Research Center, National Chengchi University.
5. Chen S-H, Huang Y-C (2005) Risk Preference and Survival Dynamics. In: Terano T, Kita H, Kaneda T, Arai K and Deghchi H (eds), Agent-Based Simulation: From Modeling Methodologies to Real-World Applications. Springer, 135–143.
6. Constantinides G-M, Donaldson J-B, Mehra R (2002) Junior Can't Borrow: A New Perspective on the Equity Premium Puzzle. Quarterly Journal of Economics 117: 269-297.
7. Chatterjee S, Giuliano P, Turnovsky S-J (2004) Capital Income Taxes and Growth in a Stochastic Economy: A Numerical Analysis of the Role of Risk Aversion and Intertemporal Substitution. Journal of Public Economic Theory 6: 277-310.
8. Epstein L-G, Zin S-E (1991) Substitution, Risk Aversion, and the Temporal Behavior of Consumption and Asset Returns: An Empirical Analysis. Journal of Political Economy 99: 263-286.
9. Friend I, Blume M-E (1975) The Demand for Risky Assets. American Economic Review 65: 900-922.
10. Gordon M-J, Paradis G-E, Rorke C-H (1972) Experimental Evidence on Alternative Portfolio Decision Rules. American Economic Review 62: 107-118.
11. Hansen L-P, Singleton K (1982) Generalized Instrumental Variables Estimation of Nonlinear Rational Expectation Models. Econometrica 50: 1269-1286.
12. Holland J, Miller J (1991) Artificial Adaptive Agents in Economic Theory. American Economic Review 81 (2): 365-370.
13. Jorion P, Giovannini A (1993) Time Series Tests of a Non-expected-Utility Model of Asset Pricing. European Economic Review 37: 1083-1100.
14. Kandel S, Stambaugh R-F (1991) Asset Returns and Intertemporal Preferences. Journal of Monetary Economics 27: 39-71.
15. Lucas D (1994) Asset Pricing with Undiversifiable Risk and Short Sales Constraints: Deepening the Equity Premium Puzzle. Journal of Monetary Economics 34: 325-341.
16. Mankiw G-N, Rotemberg J-J, Summers L-H (1985) Intertemporal Substitution in Macroeconomics. Quarterly Journal of Economics 100: 225-251.
17. Obstfeld M (1994) Risk Taking, Global Diversification, and Growth. American Economic Review 84: 1310-1329.
18. Sandroni A (2000) Do Markets Favor Agents Able to Make Accurate Predictions? Econometrica 68: 1303–1341.
19. Sciubba E (1999) The Evolution of Portfolio Rules and the Capital Asset Pricing Model. DAE Working Paper No. 9909, University of Cambridge.
20. Tesfatsion L (2001) Introduction to the Special Issue on Agent-Based Computational Economics. Journal of Economic Dynamics and Control 25: 281–293.
21. Zeldes S-P (1989) Consumption and Liquidity Constraints: An Empirical Investigation. Journal of Political Economy 97 (2): 305-346.

Explanation of binarized time series by a behavioral economic approach

Takashi Yamada* and Kazuhiro Ueda

Department of General Systems Studies,
Graduate School of Arts and Sciences, University of Tokyo,
3-8-1 Komaba, Meguro-ku, Tokyo 153-8902, Japan
tyamada@trn.dis.titech.ac.jp ueda@gregorio.c.u-tokyo.ac.jp

Abstract. The aim of this paper is to reveal the relations between time scales and time series properties by concentrating on information requisite for speculators using a genetic learning model of investor sentiment. For this purpose, first we identify the conditions to describe investor sentiment using a variety of parameters of genetic algorithm. Then we calculate auto-correlations and conditional probabilities using the estimated models in the first step. Our results show that both the amount and quality of information for the agents determine the time series properties. This implies that the preciseness of information which speculators permit depends on their time scales.

1 Introduction

The development of computer technologies and the advent of econophysics have been motivations to analyze high frequency data and, as a result, another time series properties have been successfully revealed (e.g. [6–8, 15, 27]). Besides, some studies have shown that even an appropriate method may enable us to make more profits [21, 23]. It is not easy, however, to identify the reasons or causes of such distinctions from historical data, since those data do not tell any information about the behaviors of speculators. Therefore one has to employ other approaches to clarify the mechanisms of time series data.

Several questionnaire studies have focused on the relations between time scales of speculators and information they need, and consequently have shown that there is a difference between traders at shorter horizons and investors at longer horizons [16, 20, 22, 28]. According to them, the former participants put more weight on technical trading, and the latter ones, in contrast, think fundamental information as important.

A multi-agent approach is another feasible way to explain micro-macro relations and some of the earlier studies have shown time series properties. One can fall these studies into three groups in accordance with experimental setups and the methods of time series analyses; First, the artificial markets in which

* Now at Department of Computational Intelligence and System Sciences, Interdisciplinary Graduate School of Science and Engineering, Tokyo Institute of Technology

the agents can make use of the external information can generate time series data at daily or longer time scale [1, 10, 12]. Second, in the models based on statistical physics, the agents can perceive mainly the information with respect to price/rate and therefore the data generated are usually treated as intraday ones [9, 25]. Third, well-known fundamentalist-chartist models can generate both the daily and the high frequency data [17–19], but this does not imply that the time series properties hinge on the number of fundamentalists (or chartists). In addition, it seems that few studies have taken into account the empirical fact that the information requisite for speculatros depends on their time scales.

Moreover, the recent development of behavioral economics has enabled to offer descriptive models with respect to the behaviors of speculators. Besides, some models are proposed from the experimental economic points of view [11, 30], or by applying some evidences of behavioral finance for agent-based computational economic studies [3, 14].

Therefore, the aim of this paper is to show how the relations between information and time scales affect the time series properties by an agent-based computational economic model with a behavioral economic model. For these purposes, we at first investigated whether genetic algorithm learning (hereafter GAL), which has been often used as a method of agents' learning [1, 13, 24], could represent investor sentiment developed by Barberis et al. (A model of investor sentiment: hereafter MIS) [2]. In this literature, we explored the conditions from the agents' viewpoints towards a market which were obtained when some series of typical asset-returns were given. Especially, we identified the relations between the settings of fitness calculations, namely which information the agents needed, and the possibilities to describe MIS. Then we generated sample paths using the variables estimated in the first step and showed how time to interpret information for the agents, i.e. time scale, influenced time series properties by conducting two kinds of time series analyses, auto-correlations and conditional probabilities, and by comparing between simulation results and market data.

The rest of this paper is organized as follows. The next section introduces a summary of MIS, and shows some conditions requisite for GAL to describe MIS. Then, in section 3, we explored the time series properties of generated sample paths using estimated variables of MIS. And finally, section 4 and 5 present discussion and conclusion respectively.

2 Description of MIS by GAL

Barberis et al. have developed their model of investor sentiment shown in Table 2 and 1 in order to describe market participants' "conservatism" [5] and "representativeness heuristic" [29]. More concretely, the constitutions of MIS are twofold; First, a market is either in a stable state or in an unstable one. If the market is in a stable condition, the next price movement is likely the same as the current one. While if the market is in an unstable state, the price movments is likely opposite to the current one (Table 1). The parameter λ_1 is the probability of transition from unstable condition to stable one, while λ_2 is the one from stable

Table 1. Transition probability of MIS

a. Unstable condition

	$\Delta S_{t+1} = 1$	$\Delta S_{t+1} = -1$
$\Delta S_t = 1$	π_L (< 0.5)	$1 - \pi_L$
$\Delta S_t = -1$	$1 - \pi_L$	π_L

ΔS_t: Price movement at time t

b. Stable condition

	$\Delta S_{t+1} = 1$	$\Delta S_{t+1} = -1$
$\Delta S_t = 1$	π_H (> 0.5)	$1 - \pi_H$
$\Delta S_t = -1$	$1 - \pi_H$	π_H

Table 2. An abstract of investor's recognition

$t \backslash t+1$	Stable condition	Unstable condition
Stable condition	$1 - \lambda_2$	λ_2
Unstable condition	λ_1	$1 - \lambda_1$

condition to unstable one (Table 2). Moreover, Barberis et al. postulate that the sum of λ_1 and λ_2 is less than unity and that λ_1 is smaller than λ_2. Second, the price movement in the economy is either $+1$ or -1. Therefore q_t, the probability that the market is unstable, is renewed by equation (1) in case that the asset return is the different from the previous one, or by (2) otherwise.

$$q_{t+1} = \frac{((1-\lambda_1)q_t + \lambda_2(1-q_t))(1-\pi_L)}{((1-\lambda_1)q_t + \lambda_2(1-q_t))(1-\pi_L) + (\lambda_1 q_t + (1-\lambda_2)(1-q_t))(1-\pi_H)} \tag{1}$$

$$q_{t+1} = \frac{((1-\lambda_1)q_t + \lambda_2(1-q_t))\pi_L}{((1-\lambda_1)q_t + \lambda_2(1-q_t))\pi_L + (\lambda_1 q_t + (1-\lambda_2)(1-q_t))\pi_H} \tag{2}$$

In order to define conditions requisite for GAL to describe MIS, we fed two kinds of price movements ($-+---+++$ and $+-$) to agents. Since the former series has at least the minimum information with respect to diversity, this is used to calculate the variables of MIS. While the latter series is to know how many agents share a common view to the market.

An agent has a variable, *judge*, to judge the market condition and four variables, $stable_+$, $stable_-$, $unstable_+$, and $unstable_-$, to forecast the next price movement. Each of them is a binary bit and has the following meanings:

- *judge*
  If this variable is 1 then an agent considers the market to be stable. On the other hand, if that is 0, she does the market to be unstable.
- $stable_+$, $stable_-$, $unstable_+$, $unstable_-$
  One of these variables is selected to predict the next price movement after judging the market condition by *judge*. For example, the variable $stable_+$ is used if the previous movement is $+1$ and the variable *judge* of an agent is 1. Each of them takes 1 or 0. If the variable is 1, then an agent thinks that the next price movement is $+1$, while if the variable is 0, then she forecasts that the next price movement is -1.

At first, an agent judges the market condition by her *judge*, then she makes a predict for the next movement based on that value. Therefore, in this study, the role of the agents is to check whether we can describe MIS by GAL.

We could test whether the decision rule *judge* was equivalent to the theoretical q_t by feeding two kinds of price movements in the following; At first, we saw the movements of the *judge* by feeding the PMS 1 and 2. If the fluctuation of the *judge* was similar to that of the q_t, then we thought that the GAL was able to represent the MIS.

Other setups for generating sample paths are as follows:

Number of agents 100
Number of sample paths 100
Length of each sample path 10000
Fitness function There are three types:

- **fa** An agent receives +1 if she predicts the price movement precisely.
- **fb** She receives +1 if she both judges the market condition and predicts the price movement precisely.
- **fc** She receives +1 if her observation is right as to only the market condition. While she can receive +3 if her expectation is right about the price movement, too.

These setups are to check how much and proper information an agent requires to describe MIS by GAL.

Learning frequency Every period or every 19-period
We attempted to see whether results differed from the learning interval (i.e. fast learning and slow learning) as LeBaron et al. [12] had done. The reason why the slow learning is every 19-period is that 19 is a prime number.[1]

Time horizon There are three types:

- **sa** The corresponding fitness value.
- **sb** Sum of the fitness values in the last 19 period.
- **sc** Total fitness values.

In this paper, time horizon means that how long the sum of the fitness values is needed in selecting parents.

Crossover (*pcross*) 0.6 or 0.8
Mutation (*pmutate*) 0.01 or 0.05

There are thus totally $72 = 3$ (Fitness calculations) $\times 2$ (Learning frequency) $\times$ 3 (Time horizon) $\times 2$ (Crossover) $\times 2$ (Mutation) kinds of simulations conducted.

We obtained the following conditions; First, the agents needed to know a market condition for their learnings. Second, the information used when the agents selected their parents must be up-to-date[2].

[1] Therefore, any prime number more than 10 will do as a slow learning interval. In other words, prime number can avoid the risk that learning is conducted at a specific input.

[2] For more details, see [31].

3 Time series properties of the estimated MISs

This section is to show the relations between the parameters of GAL and binarized time series properties by generating sample paths using estimated variables in the previous section, by applying them for several time series analyses and by comparing those results to actual daily and high-frequency data. In particular, we tried to show whether the results depended on the time to interpret information for the agents, i.e. the setups of the fitness calculations. Moreover, we conducted the same procedure using some parameter sets of MIS examined in previous studies.

3.1 Generation of sample paths

Since this simulation required only the variables of MIS, no agent existed or no parameter of genetic algorithm except learning frequencies was used.

The price movement ΔS_t was determined by equation (3) when the previous movement was +1, and (4) when the previous change was −1.

$$\Delta S_t = \begin{cases} +1, & \text{if } rnd() < q_t\pi_L + (1-q_t)\pi_H \\ -1, & \text{otherwise} \end{cases} \tag{3}$$

$$\Delta S_t = \begin{cases} +1, & \text{if } rnd() < q_t(1-\pi_L) + (1-q_t)(1-\pi_H) \\ -1, & \text{otherwise} \end{cases} \tag{4}$$

where $rnd() \in (0,\ 1)$ was uniform random number.

Other setups are as follows:

Number of sample paths 100
Total periods 10000
Initial q_t 0.5
Variables of MIS estimated ones in the previous section
Renewal of q_t in accordance with equations (1) and (2)
Renewal frequency of q_t the same as learning frequency (see section 2)
Initial price $S_0 = 0,\ \Delta S_t = S_t - S_{t-1} = 1$ or -1 $(t = 0,\ -1)$
The asset price S_t could take negative values, and ΔS_ts at $t = -1,\ 0$ were determined arbitrarily.

3.2 Relations between time scale and time series properties

We elicited the last 9000 periods from each of the generated sample paths and analyzed their distinctions in two ways: The one is time series analyses [26] and the other is a proof of predictabilities of tick data [21].

Sazuka et al. state that there are two aspects about binarized USD/JPY tick data: First, the first-order auto-correlation takes a negative value, while those at other lags are nearly zero (Table 3a). This is true for the estimated parameters such that the historical data are most fitted to the AR(4) process. Second, the first-order absolute AIC values are significantly larger than those at other lags in

Table 3. Auto-correlations

a. Actual data

High-frequency	Term	Lag= 1	Lag= 2
USD/JPY	1998, 1999	−0.350	0.050

Daily	Term	Lag= 1	Lag= 2
S&P 100	1984-2003	−0.020	−0.034
Nikkei225	1984-2000	0.002	−0.003
Gold Silver Index	1984-2003	−0.014	−0.017
USD/JPY	1971-2003	0.022	0.002
GBP/JPY	1971-2003	0.045	−0.008

b. Generated sample paths

	λ_1	λ_2	π_L	π_H	Lag= 1	Lag= 2
GAL 1*1	4.14×10^{-3}	7.14×10^{-3}	0.451	0.510	−0.053 (0.014)	0.007 (0.011)
GAL 2*2	4.03×10^{-3}	4.75×10^{-3}	0.462	0.523	−0.022 (0.017)	0.000 (0.010)
GAL 3*3	5.09×10^{-3}	12.63×10^{-3}	0.406	0.509	−0.129 (0.016)	0.025 (0.011)
GAL 4*4	7.96×10^{-3}	10.06×10^{-3}	0.440	0.532	−0.040 (0.015)	0.008 (0.010)
GAL 5*5	6.50×10^{-3}	16.68×10^{-3}	0.403	0.501	−0.135 (0.019)	0.051 (0.011)
GAL 6*6	4.72×10^{-2}	5.53×10^{-2}	0.398	0.540	−0.073 (0.015)	0.009 (0.010)

*1: crossover: 0.8, mutation: 0.01, LF: every period, fitness: fb
*2: crossover: 0.8, mutation: 0.01, LF: every 19-period, fitness: fb
*3: crossover: 0.8, mutation: 0.01, LF: every period, fitness: fc
*4: crossover: 0.6, mutation: 0.01, LF: every period, fitness: fc
*5: crossover: 0.8, mutation: 0.01, LF: every 19-period, fitness: fc
*6: crossover: 0.8, mutation: 0.05, LF: every 19-period, fitness fc

the fitted AR(4) process. While Mizuno et al. depict that the consecutive price increase/decrease over 8 ticks in a row shows an existence of predictability, i.e. one can forecast the price movements with over 50% probability in that case.

None of those empirical findings, however, hold for daily data. As Table 3a also shows, the absolute values of ACF at every lag are significantly smaller than that of high-frequency data[3]. Besides, the sign at lag 1 is different from the assets. On the other hand, when it comes to conditional probabilities, we could not offer the values because of the smallest observations, sadly speaking. But predictabilities cannot be seen if we calculate the probability for every 300-tick, about forty minutes long [21].

In this study, we investigated whether time series distinctions of generated sample paths were different from parameters of genetic learning, and whether there were likenesses between those sample paths and actual data. More concretely, we calculated auto-correlations and conditional probabilities, and checked whether the time to deal with the amount and quality of information, namely fitness calculations of genetic algorithm, had something related to the actual

[3] Currency data are from http://www.oanda.com/, and others are from http://finance.yahoo.com/ .

data. Furthermore, we tried to describe the behaviors of speculators or some conditions for applying MIS for actual data.

Table 3b illustrates the mean auto-correlations with standard deviations in the parentheses at lag 1 and 2. The most apparent fact is that the ACF values hinged mostly on the differences of fitness calculations, namely the first-order ACF values in case of fitness calculation *fc* were significantly smaller than those in case of fitness calculation *fb*. Besides, the second-order ACF values in case of fitness *fb* were mostly zero. Taken in this light of empirical evidences in the questionnaire studies, we can expect that the fitness calculation *fb* is suitable to explain daily data and that the fitness *fc* is proper to represent high-frequency data. In brief, we can say the following relations: If the time scale is daily, then speculators accept perfectly correct information. This is because they have time to check whether a piece of information is reliable or not. But we could not explain why the ACFs of daily data were different from the assets. If the time scale is tick-level, on the other hand, then speculators react to brand-new information even if it is not completely correct due to the similar reason. That is to say, these aspects indicate that although the rigid fitness calculation, *fb*, had some possibilities to describe binarized daily data, the extent is not sufficient. Moreover, we did not find any feature in terms of other parameters.

Next, we confirmed that the conditional probabilities depended on fitness calculations and learning frequencies (Fig. 1). Of the two parameters of genetic algorithm, the same discussion, by and large, can be said for fitness calculations, namely the amount of information which the agents could receive determined the possibilities to influence the thoughts of market participants, and thereby predictabilities[4]. This evidence is true not only for conditional probabilities per se but also for the fluctuations of the q_t. That is to say, if a consecutive price movements can be seen in the market, the q_ts, the number of speculators who consider the market as an unstable condition decreased rapidly in case of fitness calculation *fc* (for example, compare between solid line in Figure 1a and solid line in Figure 1b). Besides, the learning frequency determined the possibilities of the emergence of predictability situations. Which means, no matter how the first-order ACFs took sufficiently negative values, less learning frequencies still made speculators observe the market condition as unstable, i.e. q_t remained over 0.5, and thereby a market sentiment that the price movement would be opposite to the previous one dominated the market (Fig. 1c)[5]. The parameter set, GAL 3, thus could show relatively similar conditional probabilities because the fast learning gave speculators more opportunities to switch a view to the market, from an unstable state to a stable one (solid line in Fig. 1b), namely when the

[4] The possible reason why some parameter sets of the estimated MIS made possible to make a prediction properly is due to the variable π_L (dashed line in Fig. 1a). We conjecture that a relatively higher π_L let the agents think that the next price movements is likely the same as the previous one.

[5] We suppose that the result was due to the learning frequency. Therefore we can confirm more plausible conditional probabilities if we generate sample paths by setting learning interval to every 7 or 8 periods.

Table 4. Variable sets of MISs and auto-correlations

	λ_1	λ_2	π_H	π_L	Lag= 1	Lag= 2
explore 1	1.40×10^{-3}	6.50×10^{-2}	0.295	0.575	−0.391	0.007
explore 2	1.71×10^{-3}	0.276	0.321	0.750	−0.306	0.010

signs of a series of price movements in 6 ticks are all the same, over half market participants become to observe the market as a stable condition.[6]

Finally, we explored parameter sets which could represent the time series properties of actual high-frequency data by genetic algorithm. To conduct this procedure, we developed binary strings representing four kinds of variables of MIS, generated sample paths and conducted time series analyses as well as in 3.1. The fitness function was sum of the absolute error of the first- and second-auto-correlations, and all the ns of the conditional probabilities. Table 4 and Figure 2 present two parameter sets most fitted. There are some substantial differences between the explored parameter sets and GAL 3: First, both the parameters, π_L and π_H, in the explored models of investor sentiment are further from 0.5 than those in GAL 3. For this reason, the explored parameter sets could successfully explain the first order auto-correlation function. Second, one can find not only that the hyperbolic functions, (1) and (2), in the models of investor sentiment in Table 4 draw lush curves but also that two intersecting points between $q_t = q_{t+1}$ and hyperbolic functions are asymmetric to the center point (0.5, 0.5). This implies that traders at shorter time horizons often observe the market as unstable and that they switch their recognition to the market quite often.

4 Discussion

From the results presented in the previous section, the following meanings of the most plausible parameter set of GAL for high-frequency data can be pointed out: First, a high probability of crossover and a low probability of mutation were likely to make the agents share a common view to the market. Second, speculators react to brand-new information in no time even if it is not completely correct since they cannot have enough time to distinguish its preciseness. Probably lower π_L and higher π_H will describe actual tick data more precisely, but there is still an open question. On the other hand, if we replicate the mechanism of daily data, the information requisite for the agents is completely correct because there exists time to evaluate information. But we could not explain why the ACFs and their signs were different from the assets.

There are unfortunately several open questions other than the characteristics of daily data. First, the estimated parameters did not replicate the ACF values

[6] The reasons why GAL 4 was not suitable to describe the actual high-frequency data are: (1) the first-order ACF was relatively larger and (2) the q_t was not a rapidly decreasing function as n.

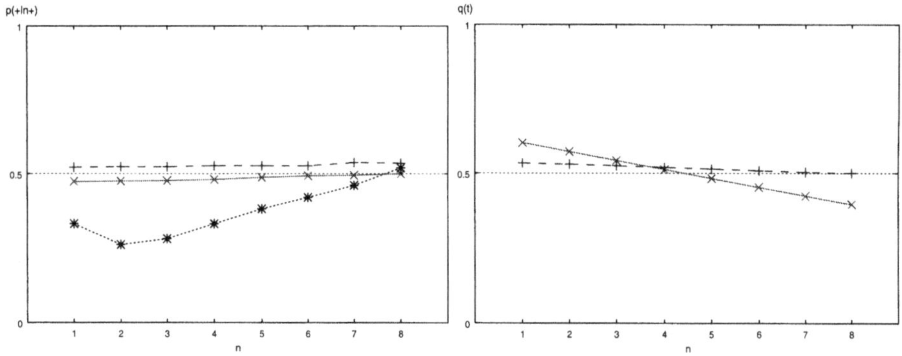

Solid line: GAL 1, Dashed line: GAL 2, Dotted line: USD/JPY tick data

a. fitness: fb

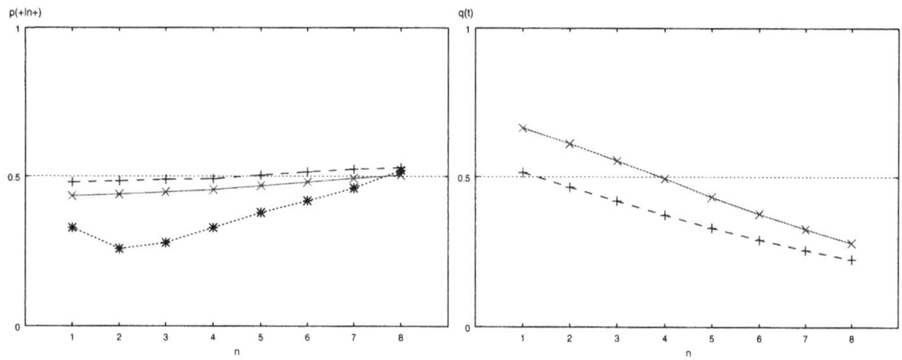

Solid line: GAL 3, Dashed line: GAL 4, Dotted line: USD/JPY tick data

b. fitness: fc (fast learning)

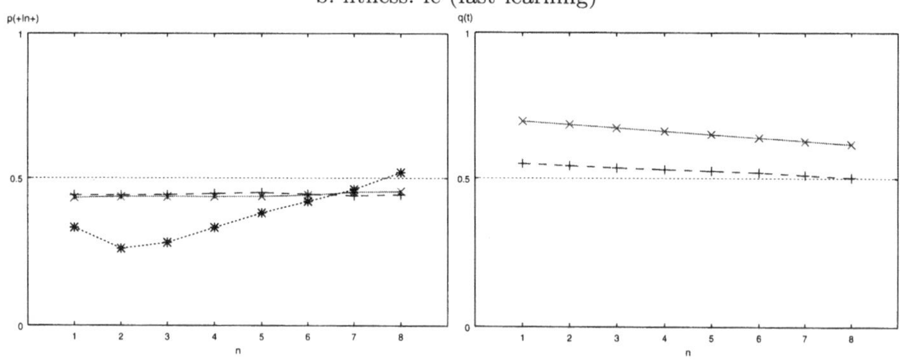

Solid line: GAL 5, Dashed line: GAL 6, Dotted line: USD/JPY

c. fitness: fc (slow learning)

Fig. 1. Differences of conditional probabilities (left panel) and q_ts (right panel)

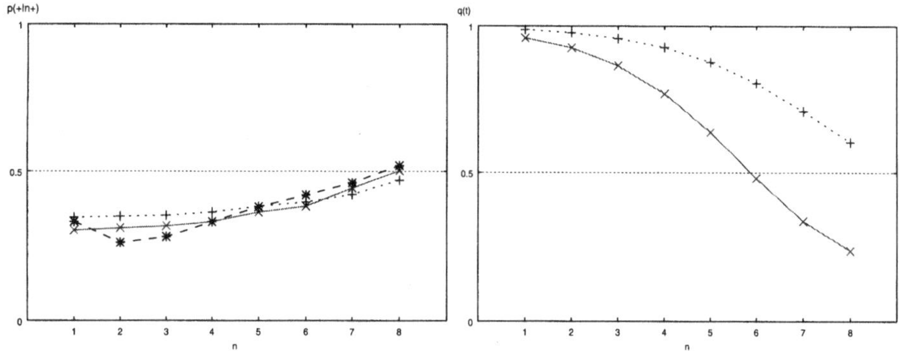

Solid line: explore 1, Dashed line: explore 2, Dotted line: USD/JPY data

Fig. 2. Differences of conditional probabilities (left panel) and q_ts (right panel)

of high-frequency data so much, even if the first-order ACF value using the most fitted parameter set, GAL3, was about half as the actual one. Second, the actual conditional probability was the smallest in case of n= 2, but any estimated MISs could not reproduce such a distinction. Therefore we suppose that a brand-new model of investor sentiment needs to be developed. Third, we have to conduct other time series analyses other than auto-correlations and conditional probabilitites. Probably, no matter how short or long the time scale may be, there will need more conditions when one re-constructs our simulation model. Finally, we have shown in section 2 that we could not obtain the varialbes of MIS when the agents selected their parents based on the sums of fitness values in the last 19-period or on the total fitness values. However, actual market participants usually mimic the behaviors of more profitable market participants in actual markets when their time scale is daily or longer. Thus one needs to know market dynamics and behaviors of market participants hinge on the way agents select their parents [4].

5 Conclusion

The aim of this paper is to show the relations between time to manage information for speculators and some stylized facts which have been captured in the previous questionnaire studies and time series analyses. To achieve this research target, we combined investor sentiment with genetic learning in an agent-based computational economic model, and conducted time series analyses using sample paths generated.

More concretely, first we explored some conditions using agents' viewpoints to the market which were obtained by feeding some series of typical asset returns to them. Two conditions should be required in order to describe the model of investor sentiment by genetic algorithm learning. First, the agents must know the market condition as their fitness values. Second, the information when the agents select their parents is always up-to-date.

Then we calculated auto-correlations and conditional probabilities of generated sample paths using the estimated variables of MIS, and compared them to the actual daily and high-frequency data. The most important finding is that time series properties are different from the way to set fitness calculations, namely the information the agents needed to know. This means if the time scale is tick-level, then speculators react to brand-new information in no time even if it is not completely correct. While if the time scale is daily or longer, then the information for speculators must be perfect since they have time to tell whether it is true or false, or to judge the power to influence the market. In brief, our results at least support the stylized facts obtained from questionnaire studies. Especially, we confirmed four possible implications to replicate high-frequency data; First, speculators are likely to have a common belief to the market. This happens when the probability of crossover is high and that of mutation is low. Second, the information which speculators were able to accept was different from the time scales. This means that the preciseness of the information was in proportion to the length of the time scale. In other words, this finding support the empirical evidences from the questionairre studies. Third, traders at shorter time horizons often observe the market as unstable. Therefore, the first-order auto-correlation takes significantly negative value. And fourth, speculators switch their recognition to the market quite often. This is partly corresponds with the first finding.

References

1. Arifovic, J., Gençay, R.: Statistical properties of genetic learning in a model of exchange rate. J. Econ. Dyn. Control **24** (2000) 981–1005
2. Barberis, N., Shleifer, A., Vishny, R.: A model of investor sentiment. J. Finan. Econ. **49** (1998) 307–343
3. Barberis, N., Huang, M., Santos, T.: Prospect theory and asset prices. Q. J. E. **116** (2001) 1–53
4. Blume, L., Easley, D.: Evolution and market behavior. J. Econ. Th. **58** (1992) 9–40
5. Edwards, W.: Conservatism in human information processing. in: Kleinmuts, B., (Eds.) Formal representation of human judgment. John Wiley & Sons (1968) 17–52
6. Goodhart, C.A.E., Ito, T., Rayne, R.: One day in June 1993: a study of the working of the Reuters 2000-2 electronic foreign exchange trading system. in: Frankel, J.A., Galli, G., Giovannini, A., (Eds.) The microstructure of foreign exchange markets. University of Chicago Press (1996) 107–182
7. Goodhart, C.A.E, O'Hara, M.: High frequency data in financial markets: issues and applications. J. Empirical Finan. **4** (1997) 73–114
8. Guillaume, D.M., Dacorogna, M.M., Dave, R., Muller, U.A., Olsen, R.B., Pictet, O.V.: From the bird's eye to the microscope: a survey of new stylized facts of the intra-daily foreign exchange markets. Finan. Stochastics **1** (1997) 95–129
9. Hirabayashi, T., Takayasu, H., Miura, H., Hamada, K.: The behavior of a threshold model of market price in stock exchange. Fractals **1** (1993) 29–40
10. Izumi, K., Ueda, K.: Phase transition in a foreign exchange market: analysis based on an artificial market approach. IEEE Trans. Evol. Comput. **5** (2001) 456–470

11. Izumi, K., Nakamura, S., Ueda, K.: Development of an artificial market model based on a field study. Info. Sci. **170** (2005) 35–63
12. LeBaron, B., Arthur, W.B., Palmer, R.: Time series properties of an artificial stock market. J. Econ. Dyn. Control **23** (1999) 1487–1516
13. LeBaron, B.: Agent-based computational finance: suggested readings and early research. J. Econ. Dyn. Control **24** (2000) 679–702
14. Levy, M., Levy, H., Solomon, S.: Microscopic simulations of financial markets: from investor behavior to market phenomena. Academic Press, San Diego, (2000)
15. Liu, Y., Gopikrishnan, P., Cizeau, P., Meyer, M., Peng, C.-K., Stanley, H.E.: The statistical properties of the volatility of price fluctuations. Phys. Rev. E **59** (1999) 1390–1423
16. Lui, Y.-H., Mole, D.: The use of fundamental and technical analyses by foreign exchange dealers: Hong Kong evidence. J. Int. Money Finan. **17** (1998) 535–545
17. Lux, T.: The socio-economic dynamics of speculative markets: interacting agents, chaos, and fat tails of return distributions. J. Econ. Behav. Org. **33** (1998) 143–165
18. Lux, T., Marchesi, M.: Scaling and criticality in a stochastic multi-agent model of a financial market. Nature **397** (1999) 498–500
19. Lux, T., Marchesi, M.: Volatility clustering in financial markets: a microsimulation of interacting agents. Int. J. Th. Appl. Finan. **3** (2000) 675–702
20. Menkhoff, L.: The noise trading approach - questionnaire evidence from foreign exchange. J. Int. Money Finan. **17** (1998) 547–564
21. Mizuno, T., Kurihara, S., Takayasu, M., Takayasu, H.: Analysis of high-resolution foreign exchange data of USD-JPY for 13 years. Phys. A **324** (2003) 296–302
22. Oberlechner, T.: Importance of technical and fundamental analysis in the European foreign exchange market. Int. J. Finan. Econ. **6** (2001) 81–93
23. Ohira, T., Sazuka, N., Marumo, K., Shimizu, T., Takayasu, M., Takayasu, H.: Predictability of currency market exchange. Phys. A **308** (2003) 1–4
24. Riechmann, T.: Genetic algorithm learning and evolutionary games. J. Econ. Dyn. Control **25** (2001) 1019–1037
25. Sato, A., Takayasu, H.: Dynamic numerical models of stock market price: from microscopic determinism to macroscopic randomness. Phys. A **250** (1998) 231–252
26. Sazuka, N., Ohira, T., Marumo, K., Shimizu, T., Takayasu, M., Takayasu, H.: A dynamical structure of high frequency currency exchange market. Phys. A **324** (2003) 366–371
27. Takayasu, H., Takayasu, M., Okazuki, M.P., Marumo, K., Shimizu, T.: Fractal properties in economics. in Novak, M.M. (eds), Paradigms of complexity. World Scientific (2000) 243–258
28. Taylor, M., Allen, H.: The use of technical analysis in the foreign exchange market. J. Int. Money Finan. **11** (1992) 304–314
29. Tversky, A., Kahneman, D.: Judgment under uncertainty: heuristics and biases. Science **185** (1974) 1124–1131
30. Ueda, K., Uchida, Y., Izumi, K., Ito, Y.: How do expert dealers make profits and reduce the risk of loss in a foreign exchange market? Proc. of the 26th annual conf. of the Cognitive Science Society, Chicago, USA (2004) 1357–1362
31. Yamada, T., Ueda, K.: Time to interpret information for speculators. mimeo.

City and Local Community

A Pedestrian Agent Model Using Relative Coordinate Systems

Toshiyuki Kaneda, Daichi Okayama

Nagoya Institute of Technology, Gokiso, Showa, Nagoya 466-8555, Japan,
kaneda.toshiyuki@nitech.ac.jp

Summary. Pedestrian agent simulation is useful for not only the analysis of an accident but also designing pedestrians' spaces. Agent Simulation of Pedestrian Flow (ASPF) has been evolved from a CA model; ASPF uses a pedestrian agent simulator with a cell space model. However, ASPF expresses an agent's position by a cell unit, which restricts the movement of a pedestrian to only four directions; this causes unnatural behaviors of pedestrians. In addition, it does not allow individual differences in the walking speed of pedestrians. This paper is to address a pedestrian agent model – ASPF ver.3 -- using a relative coordinate system by revising ASPF ver.2, and also demonstrates the special characteristics of this model by the simulation of the intersection with straight and diagonal crosswalks in front of Shibuya Station.

The pedestrian agent model proposed in this paper allows an agent to walk in any direction and also accurately simulates individual differences of walking speed. From the simulation results, it can be shown: (1) in a pedestrian agent model where multidirectional flows exist, simulated behavior closely matches actual pedestrian behavior, as demonstrated by the successful modeling of a stratification phenomenon; and (2) due to the introduction of individual differences of walking speed, a trial calculation of the percentage of remaining pedestrians has become possible.

Keyword. Pedestrian Flow, Accident Analysis, Relative Coordinate Systems

1. Introduction

Pedestrian agent simulation is useful for not only the analysis of an accident but also designing pedestrians' spaces. Agent Simulation of Pedestrian Flow (ASPF) has been evolved from a CA model; ASPF uses a pedestrian agent simulator with a cell space model. However, ASPF expresses an agent's position by a cell unit, which restricts the movement of a pedestrian to only four directions; this causes unnatural behaviors of pedestrians. In addition, it does not allow individual differences in the walking speed of pedestrians. This paper is to address a new pedestrian agent model using a relative coordinate system by revising ASPF, and also demonstrates the special characteristics of this model by the simulation of the intersection with straight and diagonal crosswalks in front of Shibuya Station.

2. Existing Pedestrian Dynamics Studies and the Feature of this Study

Pedestrian dynamics studies aim at explaining that pedestrian flow's macro phenomenon is emerged from the accumulation of the individual pedestrian's micro motive, and many existing models of the pedestrian dynamics had been reported especially on the boundary of about 2000 (Schreckenberg & Sharma, 2001). In this section we describe the feature of our model type here while comparing with the existing typical models especially focusing on the difference of continuity and discreteness on walking-space expression and agent's behavior rules and so on.(See Table 1).

Although the actual walking space is continuous, there are two categories of the model types:--the models with xy continuous coordinate system and the models that use the discrete cell-grid--, in existing models for pedestrian dynamics. Helbing et.al (1991) can be representative for the former types. His theoretic model is expressed an equation of acceleration vector for a pedestrian's walk, which is composed of the social force and the unique factors. The social force is expressed as an accelerating force, boundary repulsion power, and the interaction among other pedestrians.

In the latter types, there are cellular automata (CA) and other models that use cell spaces derived and extended from CA (for examples, Fukui &. Ishibashi, 1999, Muramatsu et.al, 1999, Burstedde et.al, 2001). Blue and Adler's model (2001) is an typical example of the cell space models. It is a tendency that the sizes of the cell of these type models are converging into 40cm each side though time slices are various in the recent researches. Further literatures are surveyed by Kaneda (2004).

Cell space models use a set of discrete condition rules for expressing a pedestrian agent's move, corresponding to others agents' locations information. This kind of cell space models is very useful, so we also adopted a cell space model in our study.

The author's laboratory had been dealing with the modeling of pedestrians' shopping-around behaviors for some years and early in the summer of 2001, we started to study the modeling of more basic pedestrian flow with the aim of trying out the performance of an agent simulation software [1]*2); in recent years the usefulness of the software has been significantly improving.

In July while designing a model, a tragic pedestrian accident occured at Akashi City Firework Festival; articles of the event and speculation of the causes were given in the press. These articles became the basis for ASPF ver.1 of the software program which was published in December; the program examined high density

[1] KK-MAS by Kozo Keikaku Engineering Inc., Japan

pedestrian flow in an L-shaped corridor and was modeled on the Asagiri Station Pedestrian Overpass (Kaneda, Yano, et.al, 2003), In January 2002, Akashi City published an accident report – ex-post facto inspection –(Akashi City, 2002) revealing further detailed information, which was then used to review and improve ASPF ver.1, reconstruct the situations of the accident and led the revision into ASPF ver.2. The simulation analysis of this program was published in February 2004 (Suzuki and Kaneda, 2004).

However, there is a fault that he/she acts fundamentally only in four directions of the orthogonalization because of the restriction of the cell grid though he/she decides the action.

In this study, in order to simulate the pedestrians' behaviors for free-angle directions, we aim at revising this model into ASPF ver.3 that have walking space model with a relative coordinate system of xy-coordinate. In this system, each agent keeps the direction of walking in his/her internal model and the conditional rules of the pedestrian behaviors that had been developed as a cell space model. This is a 'hybrid' model. These devices enable the model to simulations on multi-directions of the pedestrian flows.

Table 1. Typical models of pedestrian agent

Pedestrian Dynamics Models	Helbing (1991)	ASPF ver.3 (2005)	Blue, Adler (2001)
Space Represenation	XY-coordination system (Continuous)	XY-coordination system (Continuous)	Cell-Grid (Discrete)
Agent's Behavior Mechanisms	Vector Equations (Continuous)	Conditional Rules (Discrete)	Conditional Rules (Discrete)
Behavior Modes	Two (Ordinary, Panic)	One (Ordinary)	One (Ordinary)
Case Study Target	Mainly Two-Directions (Confronation Flows)	Multi-Directions	Mainly Two-Directions (Confronation Flows)

3. A Pedestrian Agent Model Using Relative Coordinates Systems

This model – ASPF ver.3 -- proposed in this paper is a revised ASPF ver.2 model. The spatial scale is represented by cells with 40 cm each side and the time scale is set at one step per 0.5 second. The following new agent rules are implemented:

(1) As shown in Fig.1 and Fig.2, in order to enable agents' behaviors of 360' directions, a relative coordinate system of cell space corresponding to the walking direction has been introduced for pedestrians; under this condition ASPF ver.2 pedestrian behavior rules (Fig.3) are then applied to the model. This has enabled the same pedestrian behavior rules to even be applied to pedestrians walking in different directions.

(2) It is commonly known pedestrians walk at different speeds and within a base casual distribution of average speed 1.34 m/s and a standard deviation 0.26 m/s (Henderson (1971)). Each agent is given the proper walking speeds within these parameters. Adjustments from integer to real number of forward cell movements are also made in order to represent individual differences of walking speed.

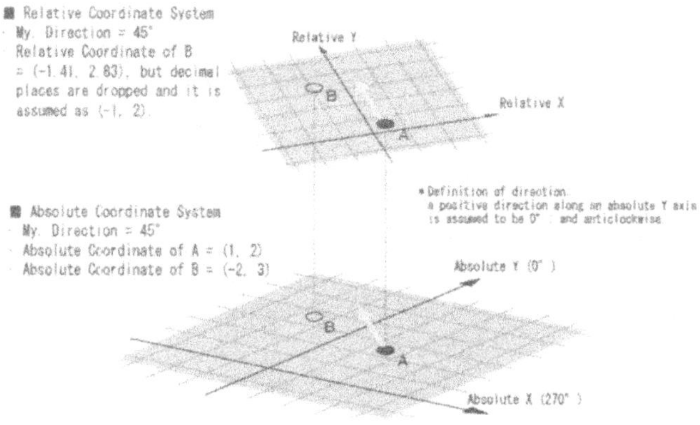

Fig. 1. Digramic Explanation of Relative Coordinate System

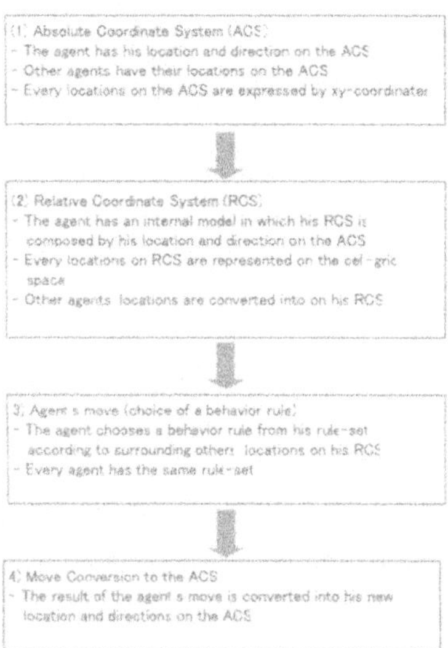

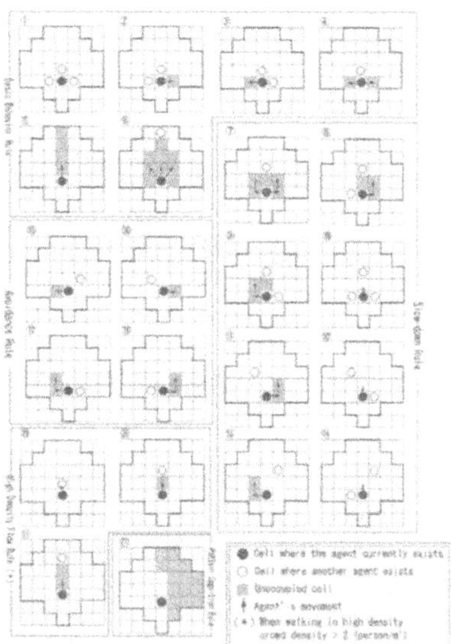

Fig. 2. Outline of Conversion between Coordinate Systems (left)
Fig. 3. Pedestrian Behavior Rules of ASPF ver.3 (right)

4. Simulation of a 'Scramble' Intersection with Straight and Diagonal Crosswalks

4.1 Target Area for the Simulation

In order to show the characteristics of the designed model, the crosswalk on a 'scramble' intersection in front of Shibuya Station was turned into a model. The 'scramble' intersection is a crossroads where traffic lights allow pedestrians to cross in any direction simultaneously. According to an on-the-spot survey, the signal cycle of the intersection is about 140 seconds. A corresponding signal control rule was introduced into the model. Fig. 4 shows a simulation algorithm. Fig. 5 shows a typical weekday evening at the Shibuya station intersection. According to the survey, at this time about 600 pedestrians are on the intersection; at an evening holiday time the number was found to be double, about 1,200 pedestrians.

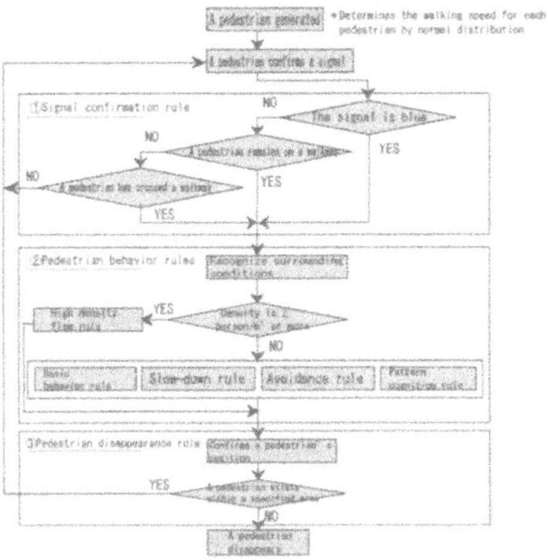

Fig. 4. Simulation Algorithm of ASPF ver.3

4.2 Simulation Outline

Based on the survey results, a weekday evening is taken to be the base case with an assumed pedestrian arrival ratio of 4.1 persons per second. The arrival ratio for a holiday evening is assumed to be double, about 8.2 persons per second. In addition, to simulate an extremely crowded situation caused by a special event for example, the arrival ratio is assumed to be 4 times the base case, giving a figure of 16.4 persons per second. The simulation focused on changes in crowd density levels and the number of pedestrians who remain at the intersection, as well as a validation check of the model. Fig. 6 shows the areas for density measurement.

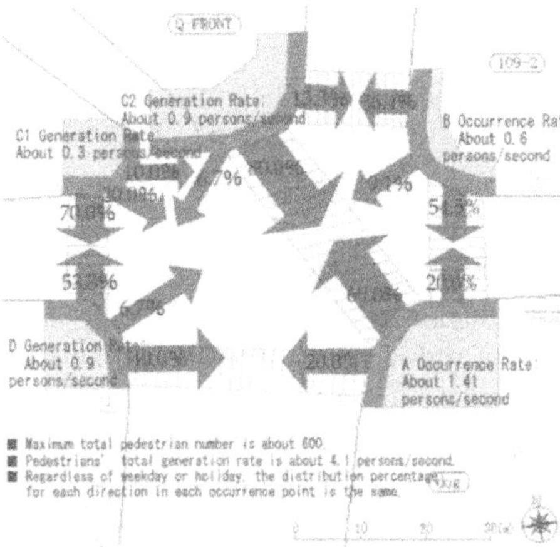

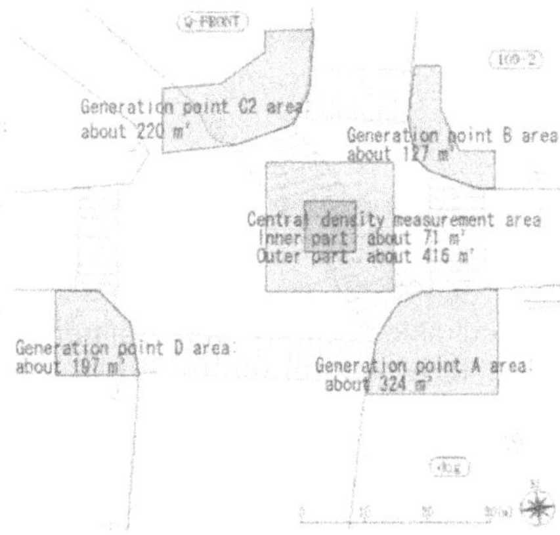

Fig. 5. Observation Results (for a Weekday Evening) (left)
Fig. 6. Density Measurement Area (right)

4.3 Results of the Simulation

Fig. 7 shows the pedestrian flow at the central part of the intersection with a holiday evening flow of double the base case. The appearance and disappearance of a stratification phenomenon between the confronting pedestrian flows can clearly be observed. Fig. 8 shows a timed change of the crowd density for each area and for the three different evenings; weekday, holiday and event cases. For each evening the maximum crowd density was measured within the inner part of the central density measurements area; for an event evening with 4 times the base case, the density substantially exceeded the critical standard of 2.0 person/m^2. In this case, the pedestrian flow speed decreases and when the pedestrian crossing signal changes from blue to red, the number of pedestrians remaining on the intersection becomes extremely large. In addition, even on a weekday and holiday evening with base case and double the base case arrival ratios, although the crowd density did not exceed 2.0 person/m^2 in any area, pedestrians still remained on the intersection. These simulation results were also observed and confirmed by the on-the-spot survey taken at the intersection. For reference, the maximum total number of pedestrians and the percentages for the remaining pedestrians on each simulated and surveyed evening are shown in Table 1.

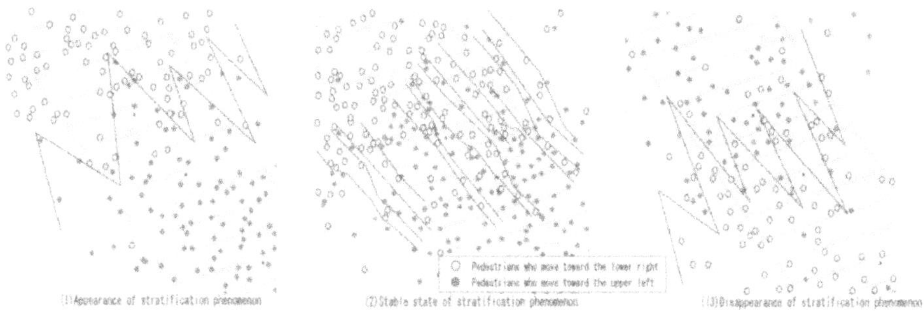

Fig. 7. Stratification Process in a Simulation

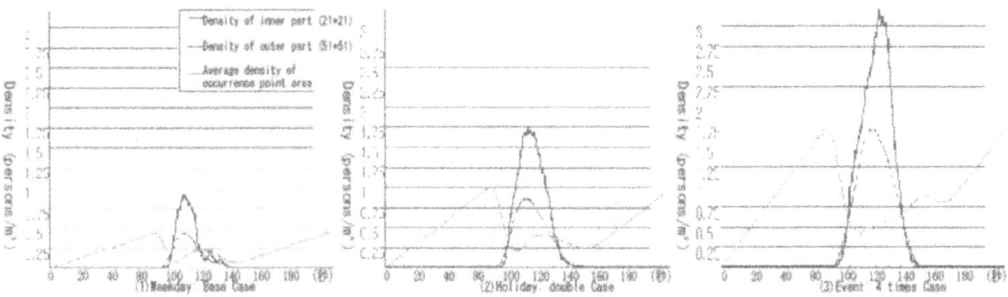

Fig. 8. Density Measurement Results

Table. 2. Percentage of Remaining Pedestrians

	Base Case	Double Case	4 times Case	Observation on weekday evening	Observation on holiday evening
Estimated number of remaining pedestrians	30	120	550	80	180
Estimated total number of pedestrians	580	1250	2250	600	1200
Percentage of remaining pedestrians (%)	5.17	9.60	24.4	13.3	15.0

4. Conclusion

The pedestrian agent model proposed in this paper allows an agent to walk in any direction and also accurately simulates individual differences of walking speed. From the simulation results, it can be shown: (1) in a pedestrian agent model where multidirectional flows exist, simulated behavior closely matches actual pedestrian behavior, as demonstrated by the successful modeling of a stratification phenomenon; and (2) due to the introduction of individual differences of walking speed, a trial calculation of the percentage of remaining pedestrians has become possible.

References

Akashi City Summer Festival Accident Investigation Committee (2002), Accident Investigation Report on the Firework Festival at the 32nd Akashi City Summer Festival (in Japanese)

Architectural Institute of Japan (2003), Complication Data for Architectural Design (Human Being), Maruzen Co., Ltd. (in Japanese)

V. J. Blue and J. L. Adler (2001), Cellular automata microsimulation for modeling bi-directional pedestrian walkways. , Transportation Research B, 35, 293/312

C. Burstedde, K. Klauck, A. Schadschneider and J. Zittarz (2001), Simulation of pedestrian dynamics using a two-dimensional cellular automaton, Physica A, 295, 507/525

M. Fukui, and Y. Ishibashi (1999), Self-Organized Phase Transitions in Cellular Automaton Models for Pedestrians, Journal of the Physical Society of Japan, 65-8, 2861/2863

D. Helbing (1991), A mathematical model for the behavior of pedestrians, Behavioral Science, 36, 298/310

L.F. Henderson (1971), The statistics of crowd fluids. Nature, 229, 381/383

T. Kaneda (2004), Agent Simulation of Pedestrian Flows, Journal of the Society of Instrument and Control Engineers, 43/12, 950-955 (in Japanese)

T. Kaneda, H. Yano, et.al (2003), A Study on Pedestrian Flow by Using an Agent Model - A Simulation Analysis on the Asagiri Overpass Accident, 2001--, T. Terano, H. Deguchi and K. Takadama (eds), Meeting the Challenge of Social Problems via Agent-Based Simulation, Springerm, 185/196

M. Muramatsu, T. Irie and T. Nagatani (1999), Jamming Transition in Pedestrian Counter flow, Physica A, 267, 487/498

A. Schadschneidar (2001), Celluar Automaton Approach to Pedestrian Dynamics-Theory, M. Schreckenberg and S. D. Sharma (eds), Pedestrian and Evacuation Dynamics, Springer-Verlag, 75/86

T. Suzuki, T. Kaneda (2004), A Simulation Analysis on Pedestrian Flow Management with an Agent-Based Approach -A Reconstruction of the Pedestrian Overpass Accident in Akashi City -, T. Terano, H. Kita, T. Kaneda and K. Arai (eds), Proceedings of The Third International Workshop on Agent- Based Approaches in Economics and Social Complex Systems (AESCS'04), 269/276

A Basic City Simulation Model for Evaluating Social Phenomena

Manabu Ichikawa[1], Yuhsuke Koyama[2], Hiroshi Deguchi[2]

Introduction

In recent years, researches which use multi agent systems become active, especially in the range of the economic phenomenon, the traffic flow simulation, and the city simulation. The research which targets the human society and social phenomenon increases. In this research, we target the infection which has a possibility to happen in every city. As a preparation for developing the simulation model of the infection, we make a basic city model using SOARS. SOARS is a new type agent based simulation framework<http://www.cs.dis.titech.ac.jp/>. In this simulation model, all agents act their dairy activity according to rules we prepare and have some interactions. From this model, we can know how the infection expands, how many agents are infected and how much infection percentage gives the city a knockdown. In this description, we analyze the expansion of the infection and show the result and the possibility of this simulation model.

A Basic City Simulation Model

In this section, we show the framework of this simulation model. The simulation step time of this model is set to 30 minutes, so all agents have an opportunity to do a decision-making 48 times a day.

[1] Department of Computational Intelligence and Systems Science, Tokyo Institute of Technology,4259 Nagatsuta-cho, Midori-ku Yokohama, 226-8502 Japan; ichikawa@cs.dis.titech.ac.jp

[2] Department of Computational Intelligence and Systems Science, Tokyo Institute of Technology, {koyama, deguchi}@dis.titech.ac.jp

A Map of a City

In this basic city, there are 5 main areas and many offices, houses, shops and restaurants and agents move here and there according to moving roles which we prepare for them. Fig.1 shows the location of 5 areas in the city.

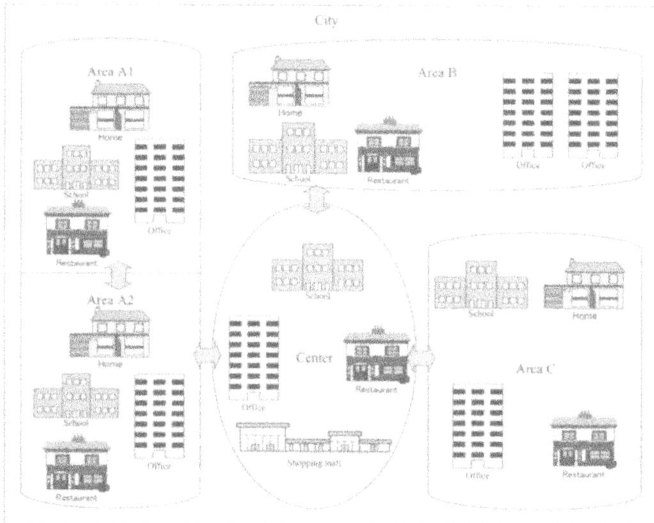

Fig. 1.The Map of the City

We explain the detail of each area in this subsection. In Area A1 and Area A2, there are 384 homes, an elementary school, a junior high school, a restaurant, a big shop, a park and some offices and 1032 agents live in these areas. We assume Area A1 and Area A2 residential areas.

In Area B, there are 245 homes, an elementary school, a junior high school, three small restaurants, a big restaurant, a big shop, a park and some offices and 508 agents live in this area. We assume Area B a business zone. In this area, there are many agents who live alone and a lot of offices exist in this area.

In Area C, there are 106 homes, an elementary school, a junior high school, a restaurant, a shop, a park and some offices and 297 agents live in this area. We assume Area C a suburb. In this area, there are many agents who live with their children and parents.

Center area which is located in the center of this city is assumed to be a downtown. There is a shopping mall, a high school, a cram school, some shops, some restaurant, and some offices and no homes in this area. No agents live in this area. As it shown in Fig. 1, all agents who want to go to the other area from some area have to go through the Center area.

The Definition of the Spot

In this simulation model, we make about 800 places where agents can move. We call each place spot in SOARS. The City spot has three properties. All of these properties are implemented as set-variables such as "area", "school" and "park". Table1 shows the detail of these set-variables.

Table 1. Properties of the city spot

set variable	Details
area	Stores all area spots in the city.
school	Stores all school spots in the city.
park	Stores all parks in the city.

All area spots, Area A1 spot, Area A2 spot, Area B spot, Area C spot and Center spot have six properties. One of these properties is implemented as keyword "area" and this keyword shows the name of the area spot. The Area A1 spot takes "A1" as a value. The others are implemented as set-variables such as "home", "office", "shop", "restaurant" and "park". These set-variables store all spots which are located in each area. Table2 shows the detail of these set-variables.

Table 2. Properties of the area spot

Property	Details
area	Shows the area of the spot.
home	Stores all home spots which are located in the area.
office	Stores all office spots which are located in the area.
shop	Stores all shop spots which are located in the area.
restaurant	Stores all restaurant spots which are located in the area.
park	Stores all park spots which are located in the area.

All other spots are thought to be a place spot where agents can stay and do some activities and also a place where a disease is infected with. All these spots have six properties. One of these properties is implemented as keyword "area" and this keyword takes the area name where the spot located in. The value of this keyword is one of the five area spots' name. Another one of these properties is implemented as set-variable "areaS". The set-variable "areaS" stores the area spot where the place spot is located in. The place spot also has a probability value to calculate infection percentage. The others are variables which are used to calculate the number of agents who are in the place spot, the number of infected agents, and the rate of infected agents and the agents who are not infected. These variables are implemented as variables "numN", "numP" and "rate". Table3 shows the detail of these properties.

Table 3. Properties of the place spot

Property	Details
area	Shows the area of the spot.
areas	Stores the area spot which the place spot located in.
p	Uses to calculate the infection percentage.
numN	Keep the number of agents who are not infected.
numP	Keep the number of agents who are infected.
rate	Keep the rate of the infected agents and the agents who are not infected.

The Definition of the Agent

There are 1837 agents in this simulation model and all agents are parted by five roles. They are “man”, “woman”, “child”, “oldman” and “oldwoman”.

The role “man” means an adult man who has a job. In this simulation model, there are 505 agents whose role is “man” and their job is whether working in the company or in the service industry.

The role “woman” means an adult woman. Some of these agents have a job and the others don’t. There are 555 agents whose role is “woman” in this model and the half of them have a job. 140 agents who live alone or live only with her children have a job and the half of the remainders also have a job. Their job is the same as the agent whose role is “man”. Agents whose role is “woman” and who don’t have a job are the housewife.

The role “child” means a child who goes to an elementary school, a junior high school, or a high school.There are 537 agents whose role is “child” and they live with their parents. The half of these agents goes to an elementary school and the half of the remainders goes to a junior high school and the others go to a high school.

The role “oldman” and “oldwoman” mean old man or woman who already retired their job. There are 115 agents whose role is “oldman” and 125 agents whose role is “oldwoman”.

All agents have seven properties. Three of them are implemented as keywords such as “work”, “go” and “condition”. Two of them are implemented as set-variable such as “home” and “area”. The others are implemented as a variable such as “level” and “p”. All these properties are used as the conditions of the agent movement rule.

The keyword “work” shows the place where an agent works. If the role of the agent is “child”, it takes school spot name as a value. If the role of an agent is “man” or “woman”, it takes work spot name as a value. The keyword “go” takes the place spot name where an agent is now going to as a value. The keyword "condition" means the condition of an agent. The value of this keyword changes according to infection rules which are described later.

The set-variables "home" and "area" store the home spot or the area spot. The set-variable "home" stores the home spot where an agent lives in this city and "area" stores the area spot where the home spot which is stored in "home" is located in.

The variable "level" shows the level of the infection. The variable "p" is used to all actions which are accompanied by the probability.

In addition to these seven properties, an agent whose role is "man" or "woman" has two more properties. They are implemented as set-variables such as "workplace" and "workArea". The set-variable "workPlace" stores the work place spot and "workArea" stores the area spot where the work place spot which is stored in "workPlace" is located in.

Table 4. Shows the detail of agent's properties.

Property	Details
work	Shows the name of the working spot.
go	Shows the place where the agent is now going.
condition	Shows the condition of the agent.
home	Stores the home spot where the agent lives in.
area	Stores the area spot where the home spot located in.
level	Shows the infection level.
p	Uses for all the action which is accompanied by the probability.
workPlace	Stores the place spot where the agent works in.
workArea	Stores the area spot where the work spot located in.

The Movement Rule of the Agent

In this subsection, we explain about the movement rule whose role is "child" simply as example. Agents whose role is "child" go to school everyday. If the value of the keyword "work" is "Eschool" which means the agent is the student of an elementary school, an agent goes out at 8:30 and arrives at the school at 9:00. The school ends at 15:00 and an agent gets out of school by 18:00 and goes home. An agent sometimes goes out to have fun at park after getting home. If the value of the keyword "work" is "JHschool" which means the agent is the student of a junior high school, an agent goes out at eight and arrives at the school at 8:30. The school ends at 17:00 and an agent gets out of school by 19:00. If the value of the keyword “work” is “Hschool” which means the agent is the student of the high school, an agent goes out at 7:30 and arrives at the school at 8:30. The school ends at 16:00 and an agent gets out of the school by 19:00. Some of the junior high school students and high school students go to cram school after school.

The Algorithm of the Infection and the recovery

The probability of infection depends on the number of the agents who are already infected and the number of the agents who are not infected in the same spot. The probability of infection of the agent is calculated from the following formula.

$$\text{Probability of infection}(\%) = \text{parcentage of infection} \times \frac{\text{number of infected agents}}{\text{number of non infected agents}} \quad (1)$$

An infection process is executed every step. When an agent is infected, the keyword “condition” changes its value to “infected”. The keyword “condition” changes its value from “infected” to “progress” next day and the variable “level” increases. The variable “level” continues to increase everyday as far as the keyword “condition” takes “progress”. The keyword “condition” changes its value from “progress” to “cure” at a certain probability. If the keyword “condition” takes “cure” as its value, the variable “level” begins to decrease. It continues to decreases everyday as far as the keyword “condition” takes “cure” and the keyword “condition” changes from “cure” to “recover” when the variable “level” reaches to 0. However, if the variable “level” reaches to 5, the keyword “condition” changes its value from “progress” to “dead”.

We explain about the algorithm of the recovery from the infection. If the agent’s variable “level” equals 2 or 4, the agent get better which means the keyword “condition” is changed to “cure” at the probability of 20%. If the agent’s variable “level” equals 3, the agent get better at the probability of 80%. While when the agent’s variable “level” reaches to 4, the agent decided to go the hospital at the probability of 66%. Once the agent goes to the hospital, the agent gets better.

The Result of Simulation

We show the result of this simulation model. We prepare 10 infected agents whose role is “child” as the initial setting. We analyzed how the infection expanded according to the infection percentage. This time, we prepare no policy into the infection. We set infection percentage to 5%, 7%, and 10% and simulate this model five times for each infection percentage.

The Period Which the Infection Goes On a Rampage

We show the period which the infection goes on a rampage in Fig.2. From this result, the higher the infection percentage is, the shorter period the infection goes on, and the lower the infection percentage is, the longer period the infection goes on. When the infection percentage is high, a short-range policy is immediately demanded and when the infection percentage is low, a long-range policy becomes necessary.

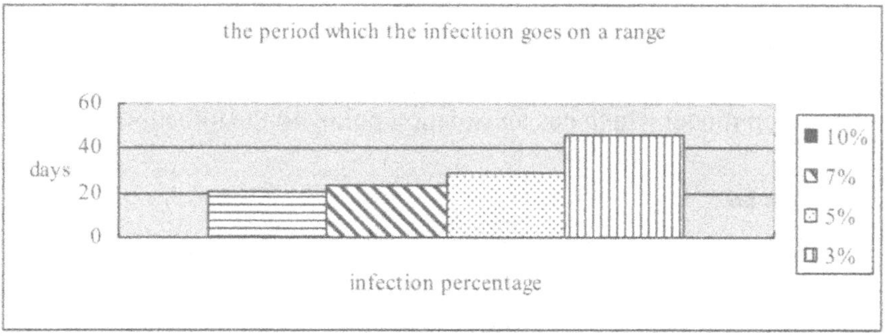

Fig. 2. The period which the infection goes on a range

The Number of The Infected Agents According To The day

We show the result of the number of infected agents according to the day in Fig.3. The higher the infection percentage is, the more rapidly the infection expanded and the lower the infection percentage is, the more slowly the infection expanded.

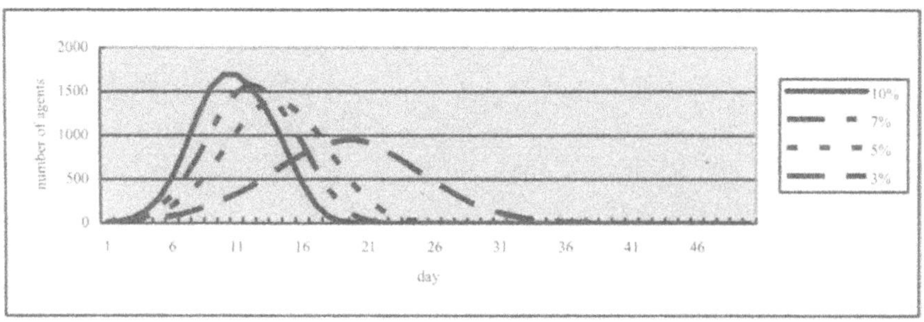

Fig. 3. The number of the infected agents according to the day

Conclusions and Future Work

In this simulation model, we prepare a basic city simulation model which includes agent's dairy movement rule and a very simple infection algorithm. From some results of this simulation model, we find what influence the infection that a policy isn't established has. From the expectation of the infection percentage and the period which the infection goes on a rampage, we know a grace to study how to cope with the situation. This simulation model keeps a possibility of being the implement of the damage expectation by the strange infection.

As a further research, to prevent the expansion of the infection we are thinking to make some policies such as the isolation policy or the distribution of the vaccine policy. If they are developed, this simulation model becomes the implement which helps us to choose an effective policy according to the situation. For the realization, we are now developing higher simulation model which has an isolation policy to the infection.

References

Hiroshi Deguchi, Hideki Tanuma, Tetsuo Shimizu (2004) SOARS: Spot Oriented Agent Role Simulator – Design and Agent Based Dynamical System. Proceedings of AESCS'04. Kyoto, Japan, pp 49-56

Manabu Ichikawa, Hideki Tanuma, Yuhsuke Koyama, Hiroshi Deguchi (2004) A Disaster Relief Simulation Model of a Building Fire. Proceedings of AIS2004, Jeju, Korea

Joshua M Epstein, Robert Axtell (1996) Growing Artificial Societies. Brookings Institution Press MIT Press

Hiroshi Deguchi (2004) Economics as an Agent-Based Complex System. Springer-Verlag, Berlin Tokyo Heidelberg New York

Analyzing Tow-Away Policy in a Local Community

Masato KOBAYASHI, Hiroshi DEGUCHI, and Takao TERANO

Department of Computational Intelligence and Systems Science
Interdisciplinary Graduate School of Science and Engineering
Tokyo Institute of Technology
4259 Nagatsuta-cho, Midori-ku Yokohama, 226-8502 Japan
masato@gssm-tsukuba.org
deguchi@dis.titech.ac.jp
terano@dis.titech.ac.jp

1 Motivation and Empirical Observation

The illegal parking in a residential area is a serious problem. Recently, they have broadcasted an interesting case at Nishidai housing area in Itabashi-ku Tokyo, Japan. They have a large marketing mall nearby with an enough large pay parking lot. However, outsiders to visit the mall tend to illegally park on a private road in the multifamily housing's site. Because the patrol of the police is difficult in the private road, the habitants have much troubles such as minor collision with inhabitants, or traffic obstructions. They must develop their own tow-away policy: without direct fines by the government, they have to form a policy group against illegal drivers.

In the United States, the police frequently patrol to prevent parking violation in the top down manner. On the other hand, in a Japanese local community where people have strong ties on a inhabitants cooperation, we do not observe terrible parking violation, because they will pay attention to the outsiders. Forming inhabitants network in the bottom up manner will be good idea to inhibit parking violation and receiving support from the local government.

Our problem in this paper is to define such tow-away policy problems and to explore the better policies using agent-based modeling methodologies. In the agent-based modeling literature, Axelrod has focused on the commitment process among agents as meta norm and analyzed it by using agent based modeling. He has reported mutual commitment among agents will construct norms in a community [1] [2]. Fixing-Broken-Windows theory also focuses on another aspect of community policy and crime-control issues [3]. From the game theoretic viewpoints, Okada have treated networking process [5].

The intermediary network is very important for problem solving in a community. We pay attention to the growing process of networking in a community.

In this paper, we assume three kinds of agents: inhabitants, outsider drivers, and the local government. Our agent-based modeling will cope with the mutual commitment, the network formation in inhabitants, and the norm formation of outsider drivers.

The simulation model consists of (1) Driver agents, who will illegally park; they make decision under the condition of a parking fee, tow-away by patrol of inhabitants, penal regulations from administration; (2) Inhabitant agents, who make decisions under the condition of a patrol cost of illegal parking, a cost of the community network participation, and a cost of psychological pressure in the community when they do not participate in the network; and (3) The local government agent, who will support the tow-away policy if the number of participants to the local network community to resist to illegal parking becomes the majority.

Under these assumptions, we investigate what kind of condition will the tow-away policy successful.

The remaining parts of the paper are summarized as follows:

In the next section we make a model for a tow-away policy. In section three several cases for simulation experiments and its parameters are given. Section four shows the result of simulation. We also analyze the result. Section five gives concluding remarks.

2 Modeling Tow-Away Policy Simulation

We model the three kinds of agents as follows:

- The drivers come from the outside into the local community. They make decisions whether to pay the parking fee or to illegally park in a residential area.When they illegally park, they get the reward of the parking fee free. But, when they are discovered by the inhabitants patrol, they must pay the cost of tow-away and a cost of the penalty fee.
- Inhabitants make two kinds of decisions. 1) whether they monitor illegal parking or not, 2) whether they participate in the inhabitants network to resist illegal parking or not.
  They must pay the cost to patrol in proportion with the number of illegal drivers. However, when the number of the network participants becomes larger than some constant value, they are supported by the local government.They also must pay the cost to participate in the network of inhabitants.The cost will decrease in proportion with the number of participant larger. When illegal parking vehicles exist, if inhabitants do not participate in the network, they will receive some negative sanction in proportion with both the number of illegally parking vehicles and the number of a network participants.

- When the number of inhabitants participating in network increases some constant value, the local government recognizes the role of the network to prevent illegal parking. There is time delay between the changes of the number of network participants and the recognition of the network by the government. We distinguish two cases where there is support for network formation or not.

We define variables and constants used in the model as follows:

α : Parking Fee (cost)
β : Negative sanction from inhabitants for illegal parking (cost)
γ : Negative sanction from administration for illegal parking (cost)
δ : Cost for patrolling illegal parking
ϵ : Administration support for patrol
ζ : Cost for network formation
η : Administration support for network formation
θ : Negative sanction as psychological pressure from local community (cost)
κ : Network participant ratio that is necessary so that administration executes support
D_f : Population ratio of the rule following drivers
D_d : Population ratio of the rule deviating drivers
$D_f + D_d = 1$
H_v : Population ratio of the inhabitants patrolling illegal parking
H_{nv} : Population ratio of the inhabitants not patrolling illegal parking
$H_v + H_{nv} = 1$
H_e : Population ratio of the inhabitants participating in network
H_{ne} : Population ratio of the inhabitants not participating in network
$H_e + H_{ne} = 1$

The model is formulated as shown in the following equations by using these variables and constants.

The utility of rule following drivers $E(f)$ and the utility of rule deviating drivers $E(d)$ are expressed in equation 1.

$$\begin{aligned} E(f) &= -\alpha \\ E(d) &= \begin{cases} -\beta H_v - \gamma & (\text{if } H_e \geq \kappa) \\ -\beta H_v & (\text{otherwise}) \end{cases} \end{aligned} \tag{1}$$

The utility of inhabitants patrolling illegal parking $E(v)$ and the utility of inhabitants not patrolling illegal parking $E(nv)$ are expressed in equation 2.

$$\begin{aligned} E(v) &= \begin{cases} -\delta D_d + \epsilon & (\text{if } H_e \geq \kappa) \\ -\delta D_d & (\text{otherwise}) \end{cases} \\ E(nv) &= 0 \end{aligned} \tag{2}$$

The utility of inhabitants participating in network $E(e)$ and the utility of inhabitants not participating in network $E(ne)$ are expressed in equation 3.

$$\begin{aligned} E(e) &= -\zeta H_{ne} + \eta \\ E(ne) &= -\theta D_d H_e \end{aligned} \tag{3}$$

3 Experimental Setup

We implemented the simulation model that is shown in the previous section. We have executed the simulation 800 times with 200 driver agents and 200 inhabitant agents, by changing the parameters.

Each has a learning mechanism implemented as reinforcement learning with Profit Sharing.. They execute reinforcement every constant episode step[7][4]. In this experiment, we fixed all the episode steps to three. We initialized all the action values equally. Therefore, when simulation starts, the probability of all actions are equal.

We assume some cases as follows and show parameter settings corresponding to each case.

3.1 Case 1

At first, we examine a basic case in which there is support from administration but penal regulations to violators are not so large. In this case, parameter settings are shown as follows.

α: 10, β: 5, γ: 10, δ: 2, ϵ: 20, ζ: 2, η: 2, θ: 4, κ: 50, Admin Action Delay: 3

3.2 Case 2

In this case, there is administration support for inhabitants network formation, but penal regulations from administration for deviators are small and inhabitants' patrol cost is large.

α: 10, β: 5, γ: 5, δ: 8, ϵ: 4, ζ: 2, η: 2, θ: 3, κ: 50, Admin Action Delay: 3

3.3 Case 3

In this case, penal regulations from administration for deviators are comparatively large, but there is no support from administration for the inhabitants network formation, and psychological pressure from a local community is small when inhabitant do not participate to the network.

α: 10, β: 5, γ: 20, δ: 4, ϵ: 20, ζ: 4, η: 0, θ: 0, κ: 50, Admin Action Delay: 3

3.4 Case 4a

In this case, just like a Case 3 there is no support from administration for inhabitants network formation, but psychological pressure from a community is large when inhabitant do not participate to the network.

α: 10, β: 5, γ: 20, δ: 4, ϵ: 20, ζ: 4, η: 0, θ: 5, κ: 50, Admin Action Delay: 3

3.5 Case 4b

This case is same as Case 3a, except that psychological pressure is larger.

θ: 50

3.6 Case 5

In this case, there is administration support for inhabitants network formation, and penal regulations from administration for deviators are large and inhabitants' patrol cost is small, but it takes much time until the administration starts the supporting action.

α: 10, β: 8, γ: 20, δ: 2, ϵ: 20, ζ: 2, η: 2, θ: 3, κ: 50, Admin Action Delay: 400

3.7 Case 6

This case is similar to Case 5, but administration start the action immediately.

α: 10, β: 8, γ: 20, δ: 2, ϵ: 20, ζ: 2, η: 2, θ: 3, κ: 50, Admin Action Delay: 3

3.8 Case 7

In this case, in addition to Case 6, penal regulations are strengthened and a parking fee is lowered.

α: 3, β: 30, γ: 40, δ: 2, ϵ: 20, ζ: 2, η: 2, θ: 3, κ: 50, Admin Action Delay: 3

4 Experiments and Discussion

The following graphs show execution results of simulation under the parameter settings which we defined in the previous section. The graphs denote the number of inhabitants patrolling illegal parking, the number of inhabitants participating in the network and the number of illegal parking.

4.1 Case 1 : Not effective support for network formation (small penal regulation)

In this case, inhabitants network is formed, and support from administration is provided, therefor number of inhabitants patrolling increases. However, because penal regulations are comparatively small, the number of illegal parking does not decrease in less than half.

4.2 Case 2 : Not effective support for network formation (large patrol cost)

In this case, Inhabitants network is formed. However, the number of inhabitants patrolling decrease because patrol cost is large. As a result, half greater or equal of all drivers select illegal parking.

4.3 Case 3 : Indifferent and no support

In this case, there is no support from administration to inhabitants network formation. Furthermore, a local community is indifference for a person who does not participate in network. Accordingly network is not formed, and support from administration for patrolling is not provided. The number of inhabitants patrolling decrease. Penal regulations to illegal parking by administration are not executed. As a result, the number of the illegal parking increase.

4.4 Case 4a : Large psychological pressure to the network formation 1

In this case, different from Case 3, the psychological pressure from a local community is large when inhabitants do not participate in the network.

Accordingly inhabitants shows a tendency to form the network though there is no support from administration for network formation. The ratio of network participants to nonparticipants is less than threshold value for administration to support, but the ratio rarely exceeds the threshold value. Because penal regulations for driver circuit are large, illegal parking decreases when a network participant exceeds threshold value at the moment. However, the number of illegal parking begins increase again during a short time because the number of inhabitants patrolling does not increase easily.

4.5 Case 4b : Large psychological pressure to the network formation 2

In this case, the psychological pressure from a local community is larger than Case 4a. Thus, the network participants often exceed threshold value, and support from administration is provided. The number of inhabitants patrolling increases. However, the number of illegal parking is only reduced to 1/3 because penal regulations are comparatively small.

4.6 Case 5 : Support for network formation + delayed support for solving problem

In this case, penal regulations are large and patrol cost is small, but the number of inhabitants patrolling decreases and the number of illegal parking increases, until administration acts. When support from administration begins, the number of inhabitants patrolling increases rapidly, and the number of illegal parking decreases.

4.7 Case 6 : Support for network formation + immediately support for solving problem

This case is same as Case 5 except that administrative reaction being fast. The number of illegal parking decreases at the beginning. But, the number of illegal parking remains in about 1/3 of the total number of drivers finally.

4.8 Case 7 : Case 6 + convenience to driver

In addition to Case 6, the number of violators decreases to almost 1/8 of the total number of drivers by making penal regulations large, and lowering a parking fee.

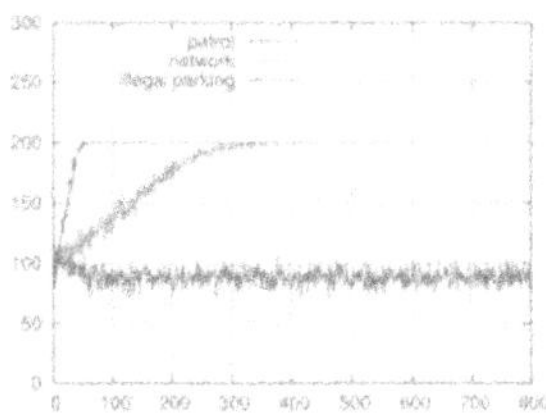

Fig. 1. Case 1

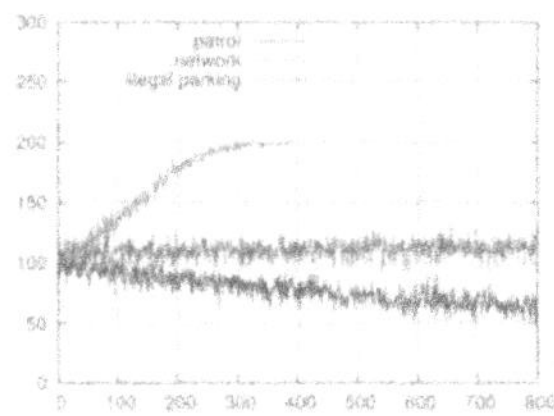

Fig. 2. Case 2

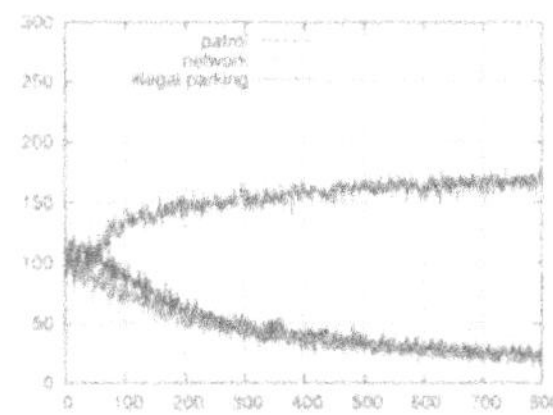

Fig. 3. Case 3

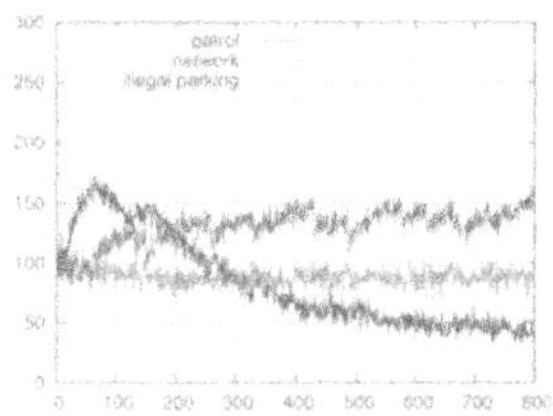

Fig. 4. Case 4a

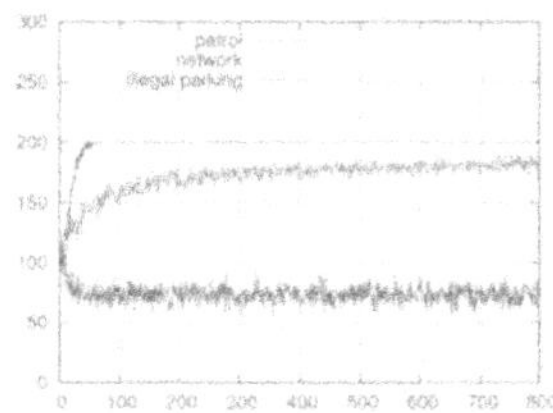

Fig. 5. Case 4b

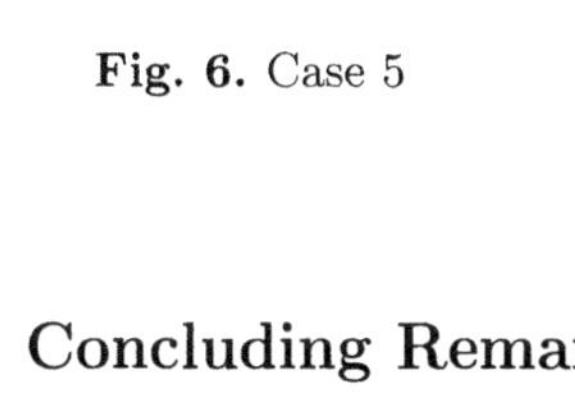

Fig. 6. Case 5

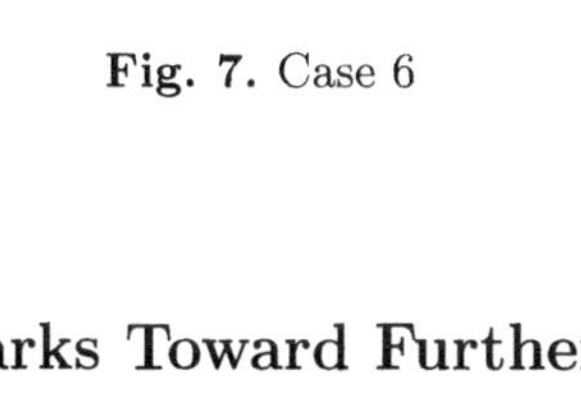

Fig. 7. Case 6

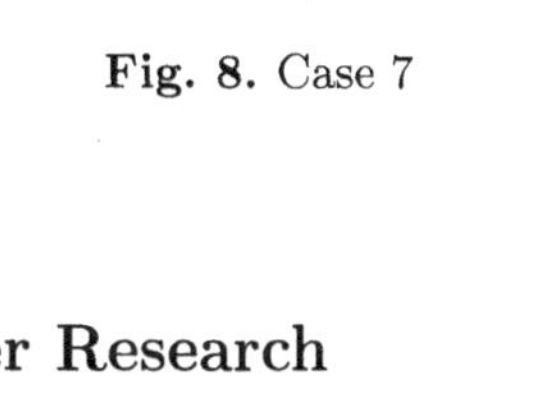

Fig. 8. Case 7

5 Concluding Remarks Toward Further Research

We have modeled a local community constituted with drivers who are going to do parking violation, inhabitants who resist parking violation and administration. We have executed the multi-agent simulation using a reinforcement learning agent. As a result patrolling illegal parking by inhabitants is insufficient to decrease illegal parking. We identify that the formation of inhabitants network, administrative support and reinforcement of penal regulations to violators are effective. Furthermore we identify that psychological pressure in a community for non-cooperators is effective to form inhabitants network.

In this model, support from administration for the community is divided into the following two types. The one is support to inhabitants forming network for problem solving. The other is direct support to problem solving. In this model, we have assumed three kinds of direct support: 1) Penal regulations to parking violators, 2) Support for inhabitants patrol, 3) Convenience to drivers (parking fee reduction).

We have extracted the cases which are not possible to solve the problem as a result because patrol cost is large or penal regulations are small though the network have formed. When the local community is indifferent and there is no support from administration for network formation, the problem solving is impossible. We have analyzed that the result depend on the administrative reaction time even if enough large network is formed. Finally, it is effective for problem solving that the support for network formation, enlarge penal regulations, support for patrol cost, lower parking fee.

On the other hand, we have divided a case of the community where problem solving is possible even if there is no support to the network formation, but psychological pressure is large. This will be near to the problem solving of traditional community. We can consider that there exists a kind of meta norm by strong social pressure for not participating in the network. However, we cannot expect strong social pressure of such mutual monitoring in modern community. It is not a pleasing situation for us.

In the modern society, it is important that inhabitants form network for problem solving, appeal to administration for the importance and draw out support. We have extracted typical ideal types for this problem by agent-based modeling.

The desirable to dismantle network after a problem has solved. The dismantling of unnecessary network is a next problem. The analysis of microscopic interaction between agents is also next problems.

References

1. Axelrod R (1997) The Complexity of Cooperation. Princeton University Press
2. Deguchi H (2004) Economics as an Agent Based Complex System. Springer-Verlag
3. Kelling GL, Coles CM (1996) Fixing Broken Windows: Restoring Order and Reducing Crime in Our Communities. Simon & Schuster
4. Miyazaki K, Kobayashi S (2001) On the Rationality of Profit Sharing in Multi-agent Reinforcement Learning. International Conference on Computational Intelligence and Multimedia Applications 2001:123–127
5. Okada A (1993) The Possibility of Cooperation in an N-Person Prisoners' Dilemma with Institutional Arrangement, Public Choice vol.77:pp.629–656
6. Randolph MN (2001) Evolution and the Capacity for Commitment, Russell Sage Foundation
7. Sutton RS, Barto AG (1998) Reinforcement Learning: An Introduction, The MIT Press

Networks

A Study on Countermeasures against Computer Virus Propagation Using an Agent-based Approach

Masayuki Ishinishi[1], Hideki Tanuma[2], and Hiroshi Deguchi[2]

[1] C4 Systems Division, Air Staff Office, Japan Defense Agency, 5-1 Ichigaya-Honmura, Shinjuku, Tokyo 162-8804, Japan, ishinishi@fw.ipsj.or.jp
[2] Interdisciplinary Graduate School of Science and Engineering, Tokyo Institute of Technology, 4259 Nagatsuta, Midori, Yokohama 226-8502, Japan, tanuma@cs.dis.titech.ac.jp, deguchi@dis.titech.ac.jp

Summary. An increase of computer viruses in recent years caused big damage on the Internet or intranet in public offices and enterprises. The damage influences at the life of the people and becomes serious problems. This paper aims to study anti-virus policies to limit damage of network system from aspects of network operation. The authors propose a diffusion model of computer viruses using agent-based approach and clarify diffusion phenomena by simulations. They also consider countermeasures against computer virus propagation by means of aforesaid method.

Key words. computer virus propagation, agent-based approach, SOARS

1 Introduction

The recent prevalence of computer viruses that infect by merely connection into the Internet has given companies and government and other public offices major damages (IPA 2005). Though some information systems have been prevented from computer virus infection by means of closed network isolated from the Internet, the epidemic occurred since an infected laptop computer was brought into such closed network. The outbreak triggered the realization of the importance of security measures.

There are many studies focused on the similarities between contagious diseases and computer viruses, the spreading of computer viruses is mainly analyzed based on the propagation model of contagious diseases (Sengoku 1996)(Okamoto 2001)(Hayashi 2004). However, There are few studies considered users, system managers, and the effect of personnel's operation and individual persons' behaviors on protection from spreading.

Thus, this research aims to examine the protective measures that minimize the damage in case of spreading computer virus infection within the Intranet of organizations mainly in operational aspects. This study reports the results from the analysis of a spreading phenomenon of a computer virus through the simulation by using the agent-based approach. The agent-based approach is suitable for description of issues in an analysis of effects on the whole system. The authors employ SOARS (Spot Oriented Agent Role Simulator), which uses the features of interactions as the simulation environment for the analysis of the spreading phenomenon of computer viruses (Deguchi 2004).

2 Simulation Process based on SOARS

SOARS equips the declarative script language, which describes the action role of agents, and the procedural action sequence control, which is based on the concept called "stage" that possesses causal sequences. In SOARS, the location in which agents move and interact is called "spot," which consists of the objects indicating the state as shown in Fig.1. The dynamic control con guration of SOARS consists of "step," which shows the unit of spreading time on simulations and "stage," which indicates the causal sequences within a step. The time is indicated in discrete time.

There are nine stages in our model. "Set-up Stage" means set-up of agents and spots. "Moving Stage" means the movement of agents between spots. "Internet Infection Stage" means the computer viral infection on the Internet. "LAN Infection Stage" means the computer viral infection in LAN. "WAN Infection Stage" means the computer viral infection in WAN. "Symptom Stage" means the destruction of data by the infected computer by the computer virus. "Detecting and Notifying Stage" means the detection of the computer viral infection and noti cation to other users and managers. "Disconnected Stage" means the disconnection from the network. "Countermeasure Stage" means the adopting of security patches.

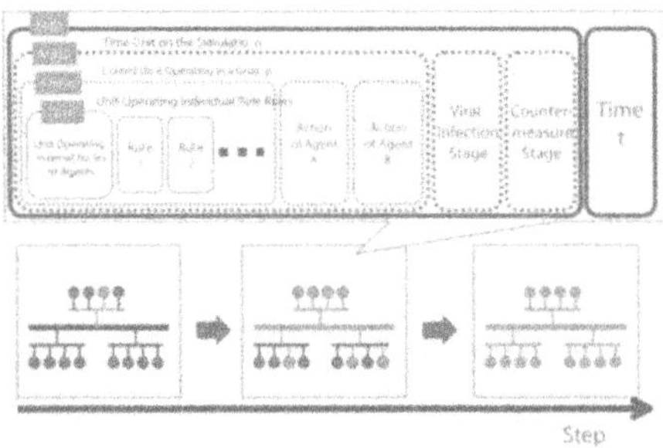

Fig. 1. Simulation Processes Based on SOARS.

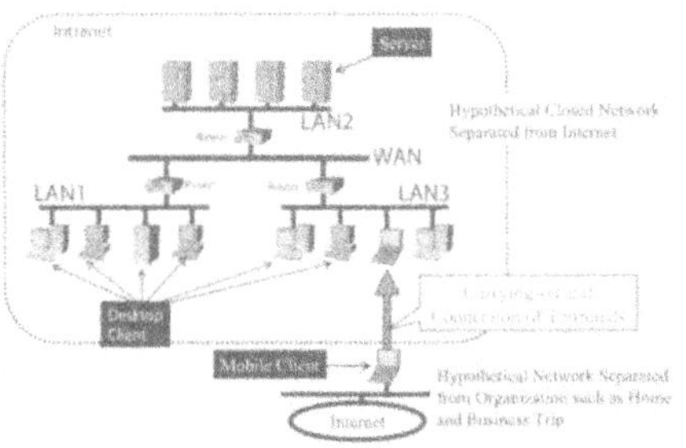

Fig. 2. Hypothetical Network.

3 Simulation Model

3.1 Modeling of Network

This research deals with the Intranet and Internet. The Intranet which consists of LAN and WAN refers to the closed environment without the Internet connection as summarized in Fig.2. The network is considered as spot, and computers are regarded as agents. The connection of computers to the network is indicated by the existence of agents in a spot. If the agent exists in the Disconnected Spot, the computer is disconnected and in stand-alone. Also, if it exists in the LAN Spot, the computer is connected to LAN. If the agent exists in the WAN Spot, the computer is connected

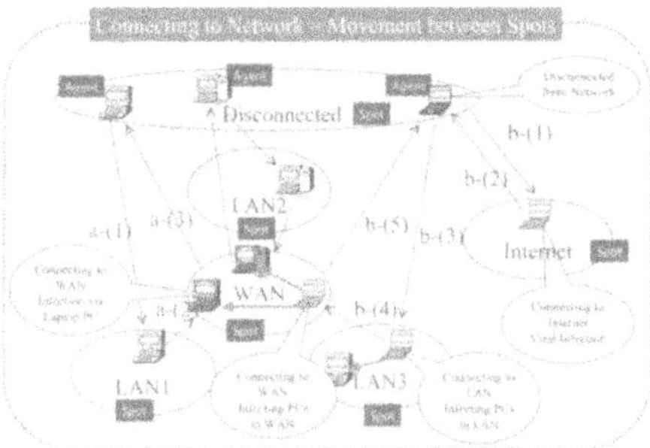

Fig. 3. Modeling of Network.

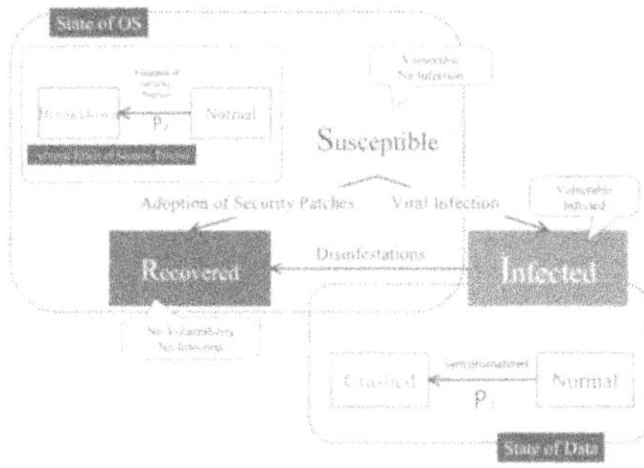

Fig. 4. Computer Virus Infection State.

to WAN. If the agent exists in the Internet Spot, the computer is connected to the Internet.

Using the spot structure of SOARS, the network structure in Fig.2 can be illustrated as that in Fig.3. For example, when user (a) connects a desktop computer to the network using WAN, exchanges data, and disconnects the network, an agent performs the following operations within the same step in the simulation: (1) it connects with the network by moving from the Disconnected Spot to the LAN Spot and exchanges data with the computers in the LAN, (2) by moving from the LAN Spot to the WAN Spot, it exchanges data with the computers in the WAN network, (3) it disconnects itself from the network by moving from the WAN Spot to the Disconnected Spot.

3.2 State of Computer

The state of computer is categorized into three: state of viral infection, state of data, and state of OS. The details are summarized in Fig.4.

In the state of computer viral infection, we employ SIR model as well as previous studies.

In terms of the state of data, there are two states: "Normal" and "Crashed." In the Symptom stage, when a computer changes from the "Susceptible" to "Infected" and the computer virus develops with probability P_1 the state changes from "Normal" to "Crashed."

The state of OS is divided into "Normal" and "BreakDown." In the Countermeasure Stage, when a computer is in "Susceptible" or "Infected" state, by adopting security patches, a breakdown of the system due to the adverse effect of the security patches takes place with probability P_2, and the state of OS changes from "Normal" to "BreakDown."

3.3 State of Computer Viruses

The computer virus with which our research deals is a computer worm, and it is infected when an infected computer by the virus connects to the network with which another computer that possesses vulnerability to the computer virus is connected. On the LAN, WAN, and Internet, a computer virus spreads from "Infected" computer to a "Susceptible" one in the Infection stages. A virus is infected to one of the "Susceptible" computers connected with the same Spot as summarized in Fig.5.

When the infecting behavior of a computer virus is illustrated with the spot orientation model, the computer virus is considered as a change in an internal attribute of an agent, namely the computer.

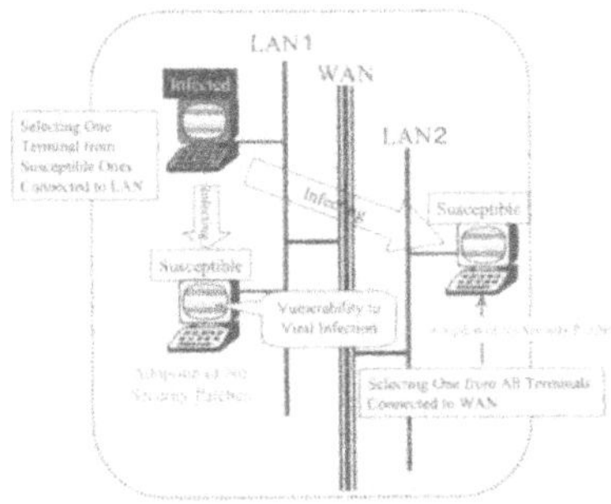

Fig. 5. Modeling of Computer Virus Infection.

4 Simulation

The simulation in this study is to analyze how a computer virus spreads over when an infected terminal is connected to a closed Intranet and what effects countermeasures by disconnecting the network and adopting security patches have on the protection of the information system.

The authors consider a network in which LAN with 100 nodes is connected to WAN as shown in Fig.2. Within the LAN nodes, there are 100 computers, and in the initial condition, it is assumed that one terminal in the whole network is infected by a computer virus.

When symptoms emerge, the probability of detection of the computer virus by the terminal user is designated as 100 % because the symptoms appear on the computer screen and they can apparently be found. Furthermore, the probability of the detection at the time of infection is set as 50 %, given the user's condition in installation of antivirus software and the user's reaction to the message from the antivirus software. Also the probability that the terminal user noti es another when having a computer with an infected or a symptom and that another user takes a countermeasure when receiving the alert is set 50 %. P_1 is set 1% and P_2 is set 0.5%.

Furthermore, the scenario for the countermeasures against the spreading of a computer virus should be considered. The scenario is divided into the following four: (1) how to detect the computer virus, (2) how a user alerts another when the terminal is infected or has a symptom, (3) how to disconnect the network to minimize the damage, and (4) which computers to adopt the extermination/security patches of the computer virus. The details are shown in Table1.

The conditions of simulation are prepared for forty major combinations of the scenarios.

Table 1. Simulation Scenario

<table>
<tr><th>Category</th><th colspan="2">Option</th><th>Note</th></tr>
<tr><td rowspan="2">Detection</td><td colspan="2">Only Terminal with Symptoms</td><td>Detect only the terminal with symptoms.</td></tr>
<tr><td colspan="2">Infected Terminal</td><td>Detect the terminal infected by a computer virus.</td></tr>
<tr><td rowspan="3">Alerting</td><td colspan="2">Neighborhood Alert</td><td>Select and alert one user in the same LAN nodes randomly.</td></tr>
<tr><td colspan="2">Alert in LAN Nodes</td><td>Alert all users in the same LAN nodes.</td></tr>
<tr><td colspan="2">Alert to All</td><td>Alert all users connected to the WAN.</td></tr>
<tr><td rowspan="4">Disconnection</td><td rowspan="3">Terminal Discon- nection</td><td>Terminal with Symptoms</td><td>Only terminal with symptoms is discon- nected and becomes in the stand-alone.</td></tr>
<tr><td>Infected Terminal</td><td>Terminal infected by a computer virus is disconnected and becomes in the stand- alone.</td></tr>
<tr><td>All Com- puters</td><td>All computers in the LAN nodes are dis- connected and become in the stand-alone.</td></tr>
<tr><td colspan="2">Nodes Disconnection</td><td>The connection point from LAN nodes to WAN is disconnected.</td></tr>
<tr><td rowspan="3">Target of Ex- termination</td><td colspan="2">Terminal with Symp- toms</td><td>Exterminate the computer virus and apply security patches to only the terminal with symptoms.</td></tr>
<tr><td colspan="2">Infected Terminal</td><td>Exterminate the computer virus and apply security patches to the computers infected by a computer virus.</td></tr>
<tr><td colspan="2">All Computers</td><td>Apply security patches to all computers.</td></tr>
</table>

5 Simulation Results

The conditions of simulation can be classi ed into six groups as a result of simulation. The rst group includes following cases: (a) to do nothing, (b) disconnection of the terminal after the detection of infected or attacked terminal, (c) disconnection of the neighbor terminal after the detection of infected or attacked terminal. The second group includes following cases: (a) extermination of the terminal after the detection of infected or attacked terminal, (b) extermination of the neighbor terminal after the detection of attacked terminals. The third group only consists of extermination of the terminal after the detection of infected or attacked terminal. The fourth group also consists of extermination of all the terminals in the LAN after the detection of infected or attacked terminals. The fth group includes following cases: (a) disconnection of all the terminals include infected or attacked terminal in the same LAN after the detection of infected or attacked terminal, (b) disconnection of all the LAN nodes after the detection of infected or attacked terminal, The six group only consists of extermination of all the terminals in WAN after the detection of infected terminals.

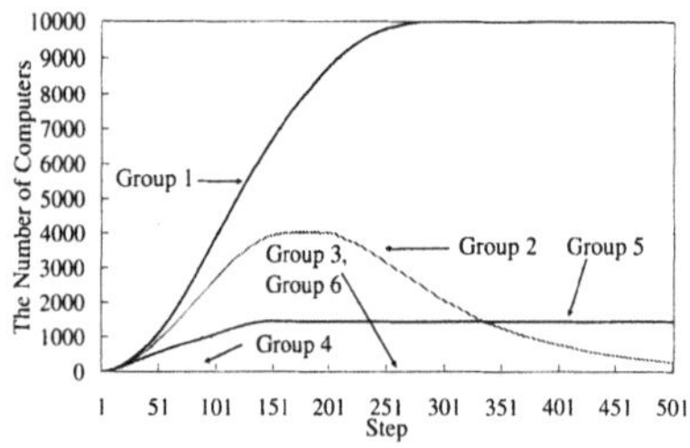

Fig. 6. The Number of Infected Computers.

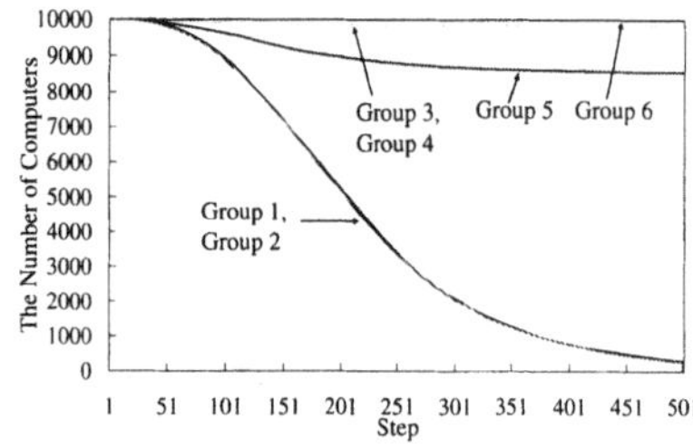

Fig. 7. The Number of Available Computers.

Fig.6 shows the change in the number of infected computers, and Fig.7 shows the change in the number of available computers; the horizontal axis represents time (step), and the vertical represents the number of the computers.

In the rst group, the number of infected computers increases, and the number of available computers decreases because the virus spreads and develops its symptoms before disconnection of terminals.

In the second group, although the number of infected computers decreases with time after an increase at the beginning, the number of available computers decreases as well as the rst group.

As a cause for this effect, it can be pointed out that the countermeasures cannot be in time for the solution because the adoption of the extermination and security patches keeps falling behind the spreading out (infection) of the computer virus, and at the time of the extermination of the virus, the expansive infection among other computers has already begun.

In the third group, however, the countermeasures are taken before the spreading of infection of the virus, and the extermination and security patches are performed before the virus develops its symptoms; the promptness can maintain the number of the available computers.

In the fourth group, the number of the available computers maintains since the countermeasures are taken before the infection and development of the symptoms of the computer virus and the infection does not spread out. This group is identi ed in case of that the user of the terminal with infection or symptoms alerts all other users in the LAN network, and this group shows even more immediate countermeasures than these in the third group whose case is that the user noti es merely neighborhood.

In fth group, the user of the terminal with infection or symptoms noti es or alerts all other users in the LAN or Computer Security Incident Response Team (CSIRT), and the LAN nodes are separated from the WAN.

In these cases, although the extermination and security patches are not performed, by disconnecting the LAN nodes, the further spreading outside the LAN with the symptoms can be stopped, and in this way, the spreading of the infection can be minimized without the extermination measure.

The sixth group includes the cases in which the user of the terminal with infection or symptoms noti es the CSIRT, and through the alert from the CSIRT, all terminal users take countermeasures such as adopting the extermination and security patches.

In these cases, whether or not the LAN nodes are separated from the WAN, all users can promptly take the countermeasures immediately after the detection of infection or symptoms, so early extermination measures are possible, and the spread and development of the symptoms of the computer virus cease with rare cases of emergence.

6 Discussion

6.1 How to Detect Infection and Send Virus Alert

In the simulation, the probability of countermeasure-taking of the users when they detect or receive the virus alert is set as 50%. This probability can be considered the extent to which the users receive the security education. From the result of the simulation, making sure to notify the managers, CSIRT etc. so that they can take the countermeasures at the organizational level is more effective to prevent the spreading of the computer virus than individual users' countermeasures.

6.2 The Effective Extermination Method for Computer Virus

The result of the simulation shows that it is difficult to protect the system merely by detecting the terminal with symptoms. Especially, the small-scale countermeasures such as alerting only the neighborhood are ineffective in exterminating the computer virus.

Moreover, as for the target of the terminal disconnection, disconnecting only the computers with infection or symptoms causes the further expansion of the computer virus because it begins spreading before being detected. Thus, a large-scale measure such as disconnecting the LAN nodes before the virus spreads into other LAN is needed.

Chen et al. note that the area in which the users take countermeasures needs to expand more quickly than the computer virus spreads in order to prevent the spreading of the computer virus (Chen 2004). More speci cally, users, when they detect a viral infection, should never take countermeasures by themselves, but instead they must notify the CSIRT so that all users can share the information and take the countermeasures at the organizational level. Needless to say, the CSIRT must direct the users to disconnect their computers immediately from the network and exterminate the computer virus when they are noti ed.

6.3 The Countermeasures for Unknown Computer Virus

As described above, a small-scale countermeasure such as alerting the neighborhood within the LAN is insufficient for the prevention from the spreading of the computer virus, and in doing so, the virus is highly likely to spread to the whole area.

Therefore, the authors propose countermeasures for an unknown virus as follows. First, when a user noti es the CSIRT immediately after detecting the infection or symptoms, the CSIRT receiving the alert directs the manager of the LAN to disconnect the connection point between the LAN and WAN to prevent the virus from spreading further. Then the manager disconnects the nodes close to the computers.

When recovering the network, to con rm the infection state of the computers, to disconnect the infected computers, and to reconnect only the networks without infection are the necessary procedures.

7 Conclusion

This research suggests the propagation model based on the agent-based approach to investigate the operational countermeasures for computer worms. It also examines, by using simulation, the effects of different countermeasures in disconnecting infected computers and user's alerting on the minimization of the damage.

The analysis of the simulation suggests effective countermeasures. These include: (1) a computer virus spreads throughout when a user detects the viral infection and takes a temporary countermeasure, (2) in order to take a countermeasure more quickly than the expansion of the virus, when the virus is detected, the global virus alert is needed, (3) immediately after the detection, the LAN nodes close to the WAN need to be disconnected rather than those close to computers.

Furthermore, the results show that it is necessary to ensure that the users notify the manager, CSIRT and alike so that they can take countermeasures at the organizational level before the users take countermeasures for their own computers.

It is also found that the global expansion of an unknown virus can be prevented by appropriate disconnecting operation of the network, even if there is no revision program for the unknown virus. Since the discussion of this research is limited to the simple network structure and computer worm, further research is needed to investigate the methods of predicting the expansion of computer viruses and effective countermeasures with the actual network structure taken into consideration.

References

Information Technology Promotion Agency, Japan (2005) Computer Virus Incident Reports, http://www.ipa.go.jp/security/english/virus/press/200504/virus200504-e.html

Sengoku,Y., Okamoto,E., Mambo,M., Uematsu,T. (1996) Analysys of Infection and Distinction of Computer Viruses in Computer Networks, 1996 International Symposium on Information Theory and Its Applications Vol.2, pp.163 166

Sengoku, Y., Okamoto, E., Mambo, M., Uematsu, T. (1996) Analysys of Infection and Distinction of Computer Viruses in Computer Networks, 1996 International Symposium on Information Theory and Its Applications, Vol.2, pp.163-166

Okamoto, T., Ishida, Y. (2001) The analysis of diffusion model of computer viruses via email, Transaction of the Institute of Electronics, Information and Communication Engineers, D-1, Vol. J84-D-I, No. 5, pp. 474-482 (in Japanese)

Hayashi, Y. (2004) Epidemic SIR dynamics on scale-free networks, Proc. of International Symposium on Dynamical Systems Theory and Its Applications to Biology and Environmental Sciences, pp. 79.

Deguchi, H., Tanuma, H., Shimizu, T. (2004) SOARS: Spot oriented agent role simulator-Design and agent based dynamical system. Proceedings of the Third International Workshop on Agent-based Approaches in Economic and Social Complex Systems (AESCS '04), pp. 49-56

Chen, L., Carley, KM. (2004) The impact of countermeasure spreading on the propagation of computer viruses. IEEE Transactions on Systems, Man and Cybernetics-Part B: Cybernetics, Vol. 34, No. 2, pp. 823-833.

Articulation of hierarchy and networks as an evolving social structure.

Juliette Rouchier, Emmanual Lazega, Lise Mounier

GREQAM-CNRS, 2 rue de la Charite, 13002 Marseille, France

juliette.rouchier@univmed.fr

Aim of the research and context

This paper describes agent-based simulations designed to study emerging networks that are created thanks to different logics of communication. The model is inspired by sociological observations led in a courthouse, in Paris, and by assumptions on the motivations for individuals to interact in this context. Two ways to choose communication partner are here represented: by following a pre-existing hierarchy; by reproducing past interactions. Both logics are described as well as their effect on the evolution of networks, and their combination is analyzed. The work is in its first phase, and although we get inspiration from real world observation, we are not leading yet a complete comparison of artificial and real society but more an exploration on how the simulations could help us in the following interview research and potentially eliminate some of our hypothesis.

The study of networks has been of high importance for social studies since some seminal papers (White et al. 1976; Granovetter 1973). Since then, the type and shape of network have been shown to be of great importance for information circulation, reciprocal obligation, and task success. One important notion that is studied in this respect is centrality of an individual in a network: the number of ties that links him or her to others. A macro view of the shape of a network can be defined, through ideas of density, cliquishness or of structural equivalence. Micro and macro levels help study the influence of the organisation on the power relations, the global behaviour for the system, its performance (Lazega and Favereau 2002).

Adding to this static approach, some people emphasize self-reinforcing dynamics for networks, and observe the evolution of ties in the system (Cowan et al. 2003), (Rouchier et al. 2000). This leads to the study of pure structural characteristics, where initial settings and evolution rules define possible emerging patterns (Jackson and Watts 2002). Both static and dynamical approaches can lead to economic analysis (performance, efficiency, innovation dissemination) or more sociological analysis (solidarity, control, regulation). Our research is sociological, focusing on the representation of dynamical processes and based on a set of data made of several static views of the system acquired through surveys led every two years (Lazega and Mounier, 2003).

An important issue in the study of networks is to show the difference between network-based dynamics and hierarchy-based dynamics (Marshack and Reichelstein 1995, Thorelli 1995). Very often they are considered to be opposed systems: for example one can study which form is most efficient in an organisation; or try to build a hierarchy "from" a network using special payoffs (Slikker and al. 2005).

The assumption here is that the relationships that are established among individuals (judges) in the Courthouse is influenced by a pre-existing formal structure that was defined regardless of the history of network. We were hence interested in deciphering the influence of emerging and non-emerging based relation, their interactions. Hence we had to consider them simultaneously. The work we have carried out is an attempt to build an abstract dynamical analysis of a network where a pre-existing hierarchy influences social exchanges and the distribution of knowledge, but where emerging properties can also be seen as relevant for individual choices of exchange partners.

Description of the court

The example that inspired our research is a court that has been studied by two of the authors following a social network analysis approach (Lazega and Mounier 2003; Lazega 2003). They wanted to witness the evolution of networks within the court, and took as an important indicator the existence of advice seeking among agents.

The court, a five century old institution, has been established to settle conflicts that arise in business relations and financial disputes in France. Because of lack of space we do not describe here the rules for evolution of population but it is sketched in the model. The characterization of individuals is made of their definition within the court (date of entry, chamber, status of president), and revealing social characteristics (age, activity, graduation year). The observed indicator is the existence or inexistence of advice seeking among any two agents, considered in general as a sign of strong interrelation among two persons. In the context of a society of peers, it is a sign of trust and respect. In a hierarchical context, it is usually a sign of recognition of the hierarchy. Following the first two surveys led in the court, it was possible to identify regularities in the forms of the relations of individuals.

First, the choice of people who are higher in the *hierarchy* has been shown to be very significant, meaning:

- Senior people, who have been in the court for longer time
- People who are president of the chamber the judge belongs to
- People who are or have been president
- People who have an official expertise in law (i.e. law degree)

The emergent way to represent "hierarchy" is to consider the most central individuals as those who are regarded by the group as the most competent. We consider them as having "high reputation" here.

Model and simulations

The artificial Court contains 147 Agents in 21 Chambers of 7. At the beginning of each year, agents change Chamber. There are at least 2 Agents who have been more than 8 years in the Court in each Chamber. Among these ancient Agents, one is chosen to be the Active President of the Chamber. Any Agent who has been President and is no longer president is a Non-Active President. Each Agent can be the owner of a law degree or not; the quantity of judges with a law degree corresponds to reality: 50%. Every year, Agents who have been in the Court more than 14 years leave and are replaced by new Agents (of age 0).

Each year is divided in 10 successive months. Agents send requests for advice to other agents every month. This is a unilateral communication since we consider that the request is the only relevant indicator in our system. A president (Active or Non-Active) makes twice less requests than a regular Judge. The number of requests made each month is a parameter of the system.

A Judge is defined by: its entry date in the Court; its Chamber; being President (Active or not); having or not having a law degree.

The only individual observation that is made on an Agent is then its centrality during the previous year, which is important for us in different respect as will be seen later. Centrality is the number of requests an Agent received from other Agents (if the same Agent asks three times, it is counted only as one request). The reason why we limit the idea of centrality to the existence or inexistence of at least one request is that in reality we do not have any data on the number of requests.

In our system, Agents always wish to seek advice from those that are higher up in the Hierarchy. As we stated in the previous section, there are two types of hierarchy, one exogenous and the other purely endogenous; and we built from it three types of rationality.

- The purely exogenous, "authority" rationality: it is based on the difference between President and non-President, and having a law degree vs. not.
- The purely endogenous one which is based on agents' centrality of the previous year, referred to as the "reputation" rationality.
- The mixed rationality is here a very simple mix of both, where Agents choose one or the other rationality alternatively (with 50% chance).

Authority rationality is defined such that:

- case 1: the Agent never was President choses its chamber President with probability ***60%,*** otherwise chooses among Presidents (active or non-active) with probabilities ***60%*** and among lawyers with ***60%***
- case 2: the Agent is or was President, then chooses with ***60%*** another President and ***60%*** a judge with a law degree

Reputation rationality is defined such that:

On the first time step: chooses randomly among more ancient Agents. Then chooses one of the 10 most central agents previous year, with probability 60 or among the others randomly.

A simulation is a succession of 15 years, each made of 10 months. In our system, Presidents make 2 requests for advice and normal Agent make 4. After the system has been initiated, Agents contact each other, leave the Court or get in, as has been explained before. Each Agent starts with an entry date randomly allocated between -13 and -1. At the end none of the Agents was there at eh beginning. The simulation is then defined by:

The type of rationality (authority, reputation, mixed, random)

Different data are observed:

- density for one year (number of links / highest possible number)

- highest and lowest centrality
- average difference in centrality and mean square deviation
- highest number of demand (several demands can be only one link)
- the classification of agents in terms of centrality, which is compared to their status as President and as a judge with a law degree.

Results

The way in which centrality is spread among agents very characteristically depends on the type of simulation. The main result is that depending on the rationality of agents, societies are very different in terms of distribution of centrality among them. Those who are most central are always presidents, but the differences in centrality is much less when hierarchy is central than when reputation works as a self-reinforcing mechanism, as can be clearly seen in Fig.1. In reputation simulation, only very few agents who have been in a tribunal for a long time concentrate most of the requests.

In a simulation with hierarchy rationality, the link between the centrality of the agent and the quality of President or of a judge with a law degree is very clear. For the hierarchy simulation the presidents are all the most central and there is a gap of centrality between those with law degree and others. When agents have mixed rationality 12 agents have a centrality of more than 94, where the others are quite equally spread from 0 to 45. It is visible that mixed rationality is an intermediate between both the other simulations in terms of centrality distribution. It is closer to reputation when it comes to the importance of the gap between the two groups of reputation. In mixed simulation, having a law degree is important for having a high centrality score, and an intermediate group appears with those agents. In Reputation simulations, only 10 agents are separated from the others by a large gap in centrality – the value of centrality for this small group is higher than 120, when the others are lower than 35.

Another type of results is important here, the density. It is the number of actual links divided by the highest possible number in this network. Values of density generated in all simulations are very far from those in reality, which is as low as 0.08. The one recorded in hierarchy simulation is the closest to this "real" value, 0.105 but is still much higher. With reputation it is 0.14 and with mixed simulation 0.13. It means that in reality few people concentrate even more requests for advice. The main reason why we might be unable to fit real data in this context may be due to the fact that in reality judges make more requests, over one year, than the 40 we allow in our simulations.

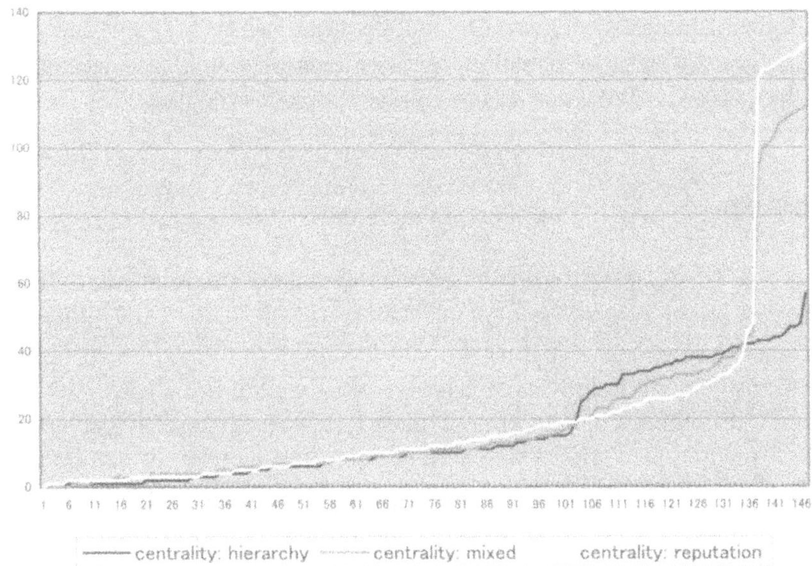

Figure 1: agents classified by centrality. X-axis is rank of agent in term of centrality and Y-axis is the value of this centrality. The three simulations are represented with reputation simulation making a clear gap in centrality, whereas hierarchy simulation displays very flat repartition of centrality.

By observing how many times centrality decreases per agent on average, it is possible to see that in hierarchy simulation the reinforcement process implying that agents with high centrality have their centrality increase is lower. With pure hierarchy agents see their centrality decrease on average 3.04 times, whereas it is only 2.38 for pure reputation and 2.48 for mixed simulations.

Discussion

These results stand for only one set of parameters in the system, which was the most plausible according to field data. The knowledge of our system is still quite superficial, although it gives promising results. Adding reputation to formal hierarchy increases the number of agents who are more central, but separates it more from the other. If we see this result in a sociological perspective, it sheds light on the importance of a formalized hierarchy for the repartition of knowledge and even power. Adding the reputation factor to the hierarchy simulations allows some agents who do not have a law degree to acquire centrality, hence reducing the tautological aspect of the hierarchy logic. At the same time the local logic and the global results are said to be more realistic by the sociologists. However, we have not tested so many values for this mixed reputation and still do not know if our results depend highly on the probabilities we stated. These tests have to be performed more thoroughly. Also, our indicators are still very complex and should be made simpler if we are to operate a systematic search on these rationalities.

Our representation of reputation and of rationality in general can be criticised. For example, there is no feed back on the success of the advice given. Also at no moment have one agent to choose between several agents following a learning process, where a form of judgement over the expertise could be involved. The process where the most contacted tend to become even more contacted over time was criticised. The fact that a highly central person (in real life) would not be have time to answer to everyone was an issue for people who listened to the description. These diverse issues are interesting from a purely abstract point of view but not really when it comes to the representation of our case study. Indeed, judges are never judged on their success, since they have more than 97% of success where making choices (a "success" in real life is the fact that the judgement is accepted right away or is confirmed by the Court of Appeal), and time is apparently not a limiting factor, since they start preparing their judgement far before the required date.

Aknowledgments

Juliette Rouchier wishes to thank Japan Society for the Promotion of Science (JSPS) for its financial support, and Toru Ishida invitation to his lab. Emmanuel Lazega and Lise Mounier wish to thank the French Ministry of Research through "Systèmes complexes en Sciences Humaines et Sociales". This work could exist thanks to Nils Ferrand for the organisation of the ACI on complex system.

References

Cowan R., Jonard N., Zimmermann J.B., 2003, The Joint Dynamics of Networks and Knowledge, in: Heterogenous Agents, Interactions and

Economic Performance, Lecture Notes in Economics and Mathematical Systems, N°521, Springer.

Jackson M. O., Watts A., 2002, The Evolution of Social and Economic Networks, Journal of Economic Theory, v. 106, iss. 2, pp. 265-95.

Granovetter M., 1973, The strength of weak ties, American Journal of Sociology, 78, 1360-1380.

Lazega E., Favereau O., 2002, Conventions and Structures in Economic Organization: Markets, Networks and Hierarchies: Introduction, Conventions and structures in economic organization: Markets, networks and hierarchies, pp. 1-28.

Lazega E., 2003, "Networks in legal organizations: On the protection of public interest in joint regulation of markets", Wiarda Chair Inaugural Address 2003, Wiarda Institute Publications, Faculty of Law, Utrecht University.

Lazega E. and Mounier L., 2003, "Interlocking Judges: On Joint External and Self-Governance of Markets", in Vincent Buskens, Werner Raub and Chris Snijders (eds), *Research in the Sociology of Organizations*, 20: 267-296.

Marschak T., Reichelstein S., 1995, Communication Requirements for Individual Agents in Networks and Hierarchies, The economics of informational decentralization: Complexity, efficiency, and stability, 1995, Kluwer Academic, pp. 311-346.

Rouchier J., Bousquet F., O'Connor M., 2001, Emergence of reputation in an artificial society: modelling non-merchant exchanges among autonomous agents, JASSS, vol 4 issue 2.

Slikker M., Gilles R.P., Norde H,Tijs S., 2005, Create hierarchies from network with equal agents at the origin, by manipulating payoff: Directed Networks, Allocation Properties and Hierarchy Formation, Mathematical Social Sciences, January 2005, vol. 49, iss. 1, pp. 55-80.

Thorelli H. B., 1995, Networks: Between Markets and Hierarchies, Integral strategy: Integration as focus, Advances in Strategic Management, vol. 11 (Part B), Greenwich, Conn. and London: JAI Press, pp. 229-247.

White, Harrison, Scott A. Boorman and Ronald L. Breiger (1976), "Social Structure From Multiple Networks I. Blockmodels of Roles and Positions", *American Journal of Sociology*, vol. 81, pp 730-80.

The socio-network model with an agent-based approach

Kousuke Shinoda[1], Yutaka Matsuo[2], Hideyuki Nakashima[3]

[1]National Defence Academy, 1-10-20 Hashirimizu, Yokosuka, Kanagawa, 239-8686, Japan
[2]National Institute of Science and Technology,
2-41-6 Aomi, Koto, Tokyo, 135-0064, Japan
[3]Future University-Hakodate, 116-2 KamedaNakanomachi, Hakodate, Hokkaido,041-8655, Japan

Abstract. In this paper, we propose a novel approach to explain the emergence of different network structures through multi-agent network simulation. Each agent as node of network has rationality and new edge to be added are chosen based on mutual common consent of all agents. The agents' rationality is to raise own position in network. That is, it is to increase its own centrality, and it votes so that its centrality is maximized. Depending on the types of centrality measures, different types of network structures are obtained. This model of growing network explains emergence of a network where many agents participate in creating it: It includes a social network where each autonomy tries to be more central, and traffic network where each region tries to be more accessible.

1 Introduction

A network is a set of items, which we call vertices, with connections between them, called edges. Around the real-world, we can see it networks. For examples, Internet, social network of acquaintance, railway[1], food webs, business relation between companies, and many others[2, 3]. Recently, there are many studies of network. These network researches focus on the consideration of large-scale statistical properties of graph, and has been driven largely by the availability of computers and communication networks that allow us together and analyze data on a scale far large than previously possible[4]. There are several models of growing network, ex, Barabási-Albert model[5], Price's model[6]. The features of their model is *preferential attachment*, and new add vertices selects to connect to which vertices, according to its preferential attachment. Their models make a network, in which degrees distribution is the power-law.

In this paper, we propose a novel approach to explain the emergence of different network structures through multi-agent network simulation. Each agent as node has rationality and new edge to be added are chosen by voting of all agents. We can see it in socio-process. These networks are constructed under a social demand, but each node aims to raise own position by adding new edges. We would like to be clear that the feature and value of our proposed model by the simulation.

2 The model of growing network

The growing network models are much attention recently. Scale-free and small-world networks are revealed to exist in many natural and artificial systems. There have been several studied on how such networks emerge, small-world or scale-free network.

In our model, each vertex cannot be freely connected to others, and all edges are chosen by voting among all vertices. An agent as vertex has the rationality that it wants to raise own position in the network. We call this approach "agent-based growing network model". In this research, we introduce the centrality of networks as the standard of agents' rationality.

2.1 The centrality of network

Centrality is a structural attribute of vertex in a network. This is a measure of how central an actor is in network, and the contribution of network position to the importance, influence, prominence of an actor in a network. This centrality has the three most commonly adopted measure of node centrality[7]: the degrees C^D, the closeness C^C, and the betweenness centrality C^B. Moreover, we also employ the pagerank C^P[8] that is used by Google search engine. This is similar to eigenvector centrality or Bonacich centrality.

1. **Degree centrality:**

   Degree is the number of own tie to others. The vertexes who have more ties become advantage position. The degree centrality of i can be defined as[7]:
   $$C_i^D = \frac{k_i}{N-1} = \frac{\sum_{j \in G} a_{ij}}{N-1}$$
   where k_i is the degree of points i. A vertex with a large degree is in direct contact to many others and being very visible is immediately recognized by others as a hub, a very active point and major channel of communication.

2. **Closeness centrality:**

   Closeness is the average of geodesic distance through the network from one vertex to another. In this case the idea that a vertex is central if it can quickly interact with all the others, not only with first neighbors. A vertex who has high closeness centrality is likely to receive information more quickly than others. The closeness centrality of point is [7]:
   $$C_i^C = (L_i)^{-1} = \frac{N-1}{\sum_{j \in G} d_{ij}}$$
   where L_i is the average geodesic distance of i. Such a measure is meaningful for connected graphs only, unless one assumes d_ij equal to a finite value, for instance the maximum possible distance $N-1$, instead of $d_{ij} = +\inf$, when there is no path between two nodes i and j.

3. **Betweenness centrality:**

   Betweenness centrality is the number of all shortest paths through the network. A vertex has high betweenness can control the flow of information, acting as a gatekeeper, and may also serve as a liaison between disparate regions of the network. This concept can be simple quantified by assuming that the communication travels just along the geodesic. The betweenness centrality of actor i can be defined as[7]:
   $$C_i^B = \frac{\sum_{j<k \in G} n_{jk}(i)/n_{jk}}{(N-1)(N-2)}$$

In the double summation at the numerator, j and k must be deferent from i. Similarly to the other centrality measures C_i^B takes on values between 0 and 1, ant it reaches its maximum when actor i falls on all geodesics.

4. **PageRank**

   PageRank[8] is the principal eigenvector of the adjacency matrix of a network to measure the relative importance of individual vertex[9]. The essential idea is that if page u has a link to page v, then the author of u is implicitly conferring some importance to page v. Let N_u be the outdegree of page u, and let $C^P(p)$ represent the importance of page p. Then the link (u, v) confers $C^P(u)/N_u$ units of rank to v. In each iteration, propagate the ranks as follows:

$$\forall_v C_{i+1}^P(v) = \sum_{u \in B_v} C_i^P(u)/N_u$$

   We continue the iterations until C^P stabilizes to within some threshold[9].

2.2 Simulation

We design an agent-based simulation to construct a network as follows:

1. **[Initial state]**
   - Denote the number of nodes n_0 in the initial state.
   - Set the vote candidate number of edge $c = \{ c \leq N(N-1)/2 \}$.
   - Set the maximum number of constructed edges.
2. **[Select candidates of new connected edge]**
   - Choose a node randomly. This node is source of a new edge.
   - Choose a destination node randomly that is not already connected from source node.
   - Add this edge to candidate list of edges. Repeat until the fill of candidate edges get c. This phase is the ended at that time if there is no edge becomes a candidate.
3. **[Evaluation of candidates and voting]**
   - For each candidate edge, each agent evaluates its value. In this case, its value is centrality of network, and agent hopes to increases own centrality by constructing new edge. Moreover, voting method adopts Borda count. This Borda count is voting system devised by Jean-Charles de Borda, used for single or multiple-seat elections[10]. With c candidate, each agent awards c to points to his first choice, $c - 1$ points to his second choice, and so on. The candidate with most points win.
4. **[connect new edge]**
   - The candidate edge with the most points is actually constructed. Terminate if the number of edges satisfies the terminal condition $e = E_{max}$. Otherwise goto step 2.

In this simulation, the growth of network is only addition of new edges, a number of vertex is constant . This is because that the proposed model is made simple. An agent as vertex has a rationality as the centrality of networks. In this paper, we introduce the four options in terms of a centrality measures, i.e., degrees centrality C^D, closeness centrality C^C, betweenness centrality C^B, and PageRank centrality C^P.

In the next section, we show the result of network evolving process, and consider the feature of the growth processes. Beside, we make clear the feature of each centrality as seed of growth in order to control of growth of network construction.

3 The result of simulation

We simulate the four agent-based growing networks, closeness, betweenness, degree, pagerank. As a result, these networks have grown up to different topology of networks according to each centrality. These network topology has the difference features. The results of simulation that uses each centrality is shown at the following.

3.1 In case of Closeness Centrality

Fig. 1 is the process of network growth with closeness centrality. This figure shows some hub in the process of evolution. This is because most edges vote to an edge that is connected to a hub, which in turn increases the closeness of most nodes. The network is similar to a scale-free network, and is good terms of connectedness.

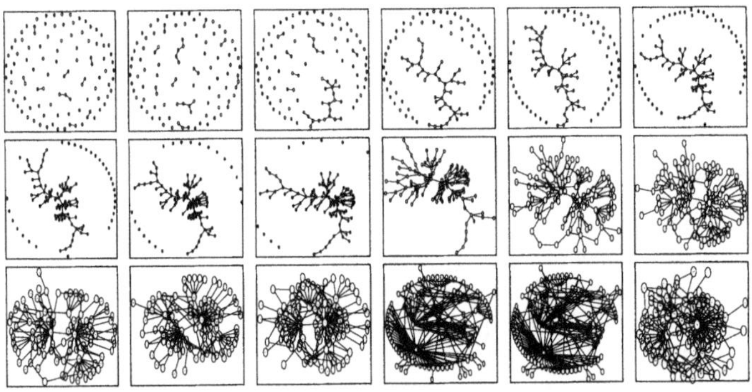

Fig. 1. An example of growth process of network with closeness evaluation(N=100)

In this case, we can see that the network topology is different depending on the size of c, case of $c \leq N$ shows no hub, but the case of $c \geq N$ has two or three hubs, especially case of $N \leq c \leq 3N$ shows Fig. 2.

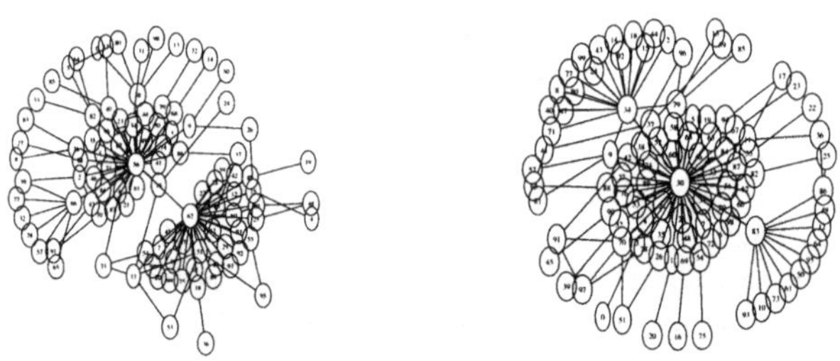

Fig. 2. The screen shot of network(N=100), which has a few hub node.

Fig. 9 shows the difference of degree distribution according to the evolving step. This figure shows a power law at the number of edge $l \leq 300 (= 3N)$.

The network consequentially obtained is similar to the real network of air-plain. This network has a few hub that is large city. Perhaps, it is guessed that the network where the cost of time and fuel doesn't change regardless of the distance, such as air-plain network, has the similar process of network evolution with closeness centrality.

3.2 In case of Degree centrality

Fig. 3 is the process of evolution with degree centrality. In case of Degree centrality, network topology grows up to a random network as addition of any one edge will only increase the degree on the both ends. Therefore, an edge is selected almost at random, but the decentralization of the degree of the node is a little, shows Fig. 10.

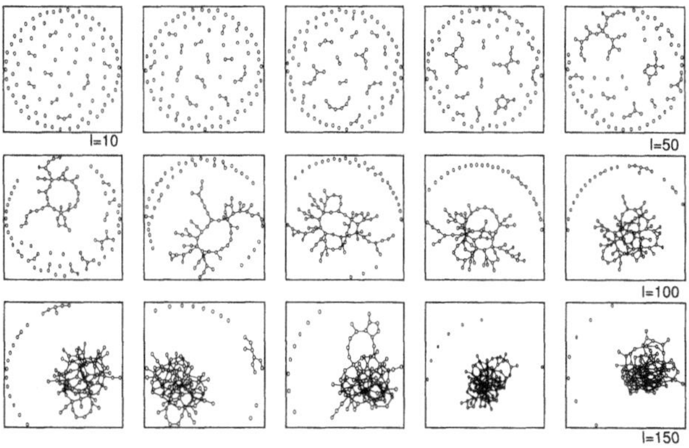

Fig. 3. The process of network growth with degrees centrality(N=100,c=200)

Fig. 4. A screenshot of network($N = 100$, $c = 200$, $r = 2.1$)

Fig. 10 shows the difference of degree distribution according to the evolving step. In case of $l \leq 100(= N)$, degree distribution shows a power law. In case of $l \geq 100(= N)$, there is no specific feature.

3.3 In case of Betweenness centrality

Fig. 5 shows a different kinds of evolution. The network topology becomes gradually connected and no hub emerges. Every edge wants to increase it betweenness, so the shortest paths become to be long. The obtained network is not efficient in terms of connectedness. Fig. 11 shows the difference of degree distribution according to growth

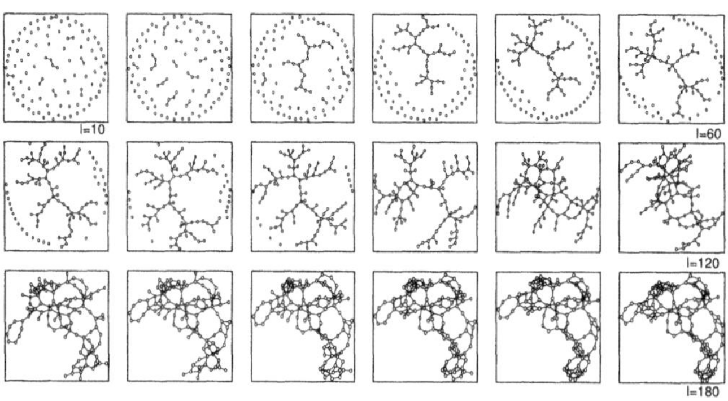

Fig. 5. The process of network growth with betweenness centrality(N=100,c=200)

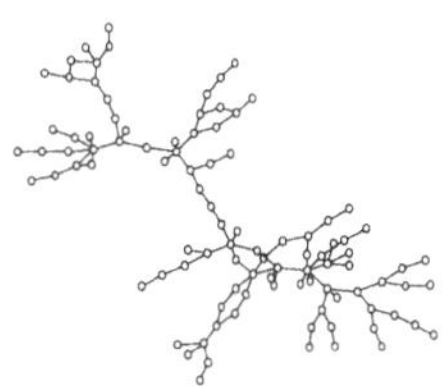

Fig. 6. A screenshot of network($N = 100$, $c = 200$, $r = 2.1$)

of network.

3.4 In case of PageRank centrality

7 is the process of evolution with PageRank centrality. In case of PageRank centrality, network topology grows up to a complete network only by a part of the vertex.

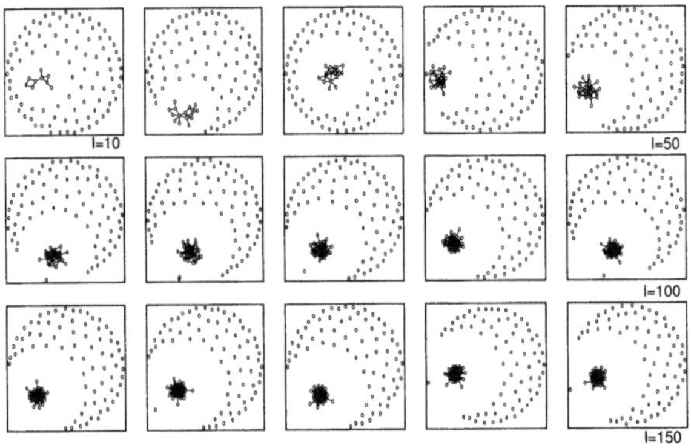

Fig. 7. An example of growth process of network with PageRank centrality(N=100)

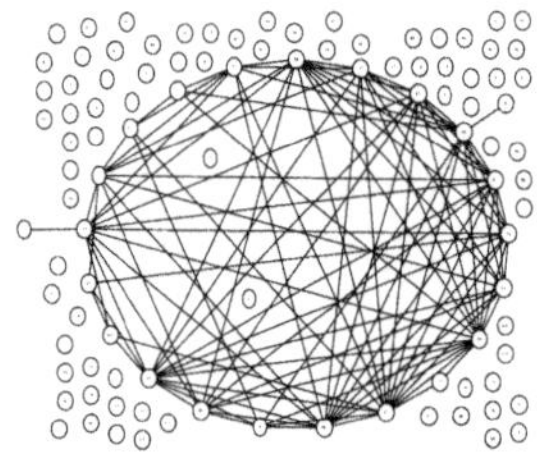

Fig. 8. A screenshot of network($N = 100$, $c = 200$, $r = 2.1$)

3.5 Degree Destribution

This behavior is difference from others in the point of network size. This composition of a complete subgraph and isolated many nodes is the characteristic of using PageRank centrality. That is, member of a community evaluate highly each other, neglecting many others.

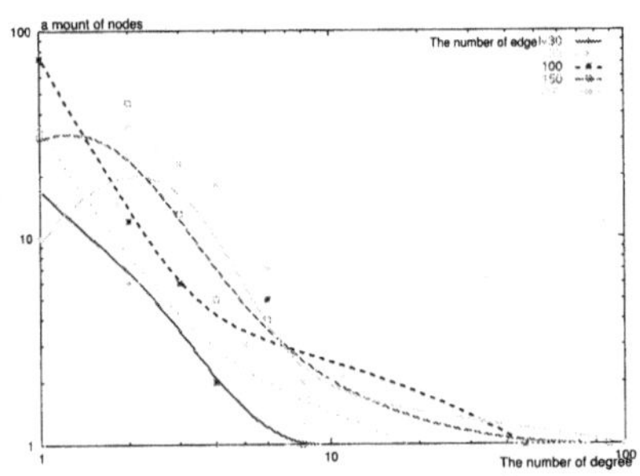

Fig. 9. closeness centrality.(N=100, c=200)

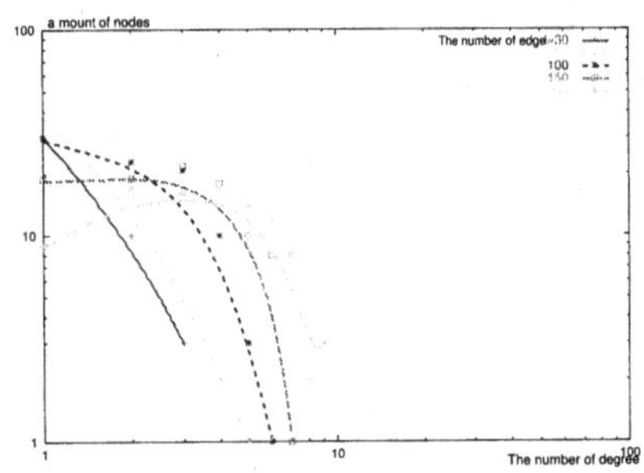

Fig. 10. degrees centrality(N=100,c=200)

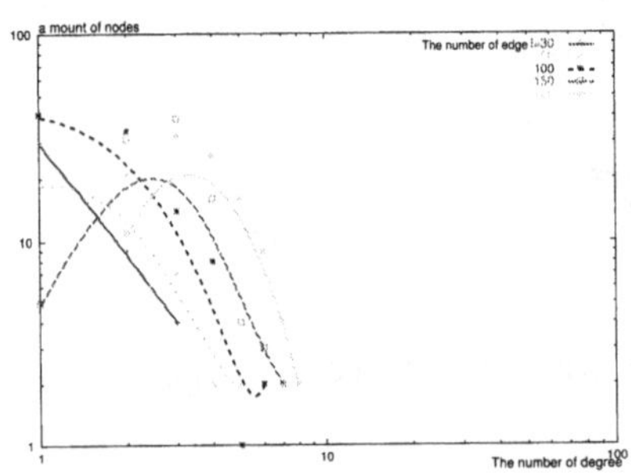

Fig. 11. betweenness centrality(N=100, c=200)

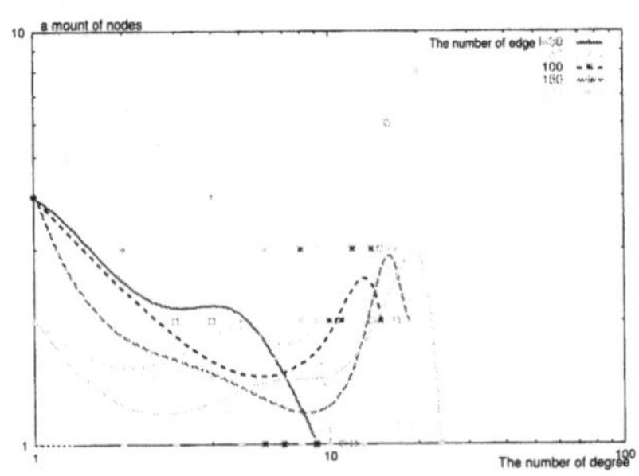

Fig. 12. PageRank.(N=100, c=200)

Fig. 12 shows the difference of degree distribution according to the evolving step. We find that a few nodes have many degrees.

3.6 The length and clustering coefficient of networks

Table 1 is the network properties L and C for each centrality. The closeness network is small length and small clustering coefficient compared with other network, and L becomes small as c becomes large size. In case of betweenness network, L is constants, but C becomes large, but smaller than closeness networks. In case of pagerank network, L is the smaller and C is the larger than others, when the isolation node is not considered.

Table 1. The defference of the properties of network, L and C, according to a number of candidate edge. This network has 100 nodes, and 105 edges (average of degrees par node $\gamma = 2.1$). The value in parentheses is numerical values include the isolation vertices

Centrality	candidate	γ	L	C
Closeness	c=20	2.1	4.54 (4.72)	0.00655 (0.00857)
Closeness	c=100	2.1	3.54 (4.54)	0.00898 (0.00766)
Closeness	c=200	2.1	3.32 (3.32)	0.00765 (0.00582)
Closeness	c=max	2.1	2.49 (2.49)	0.0212 (0.0212)
Betweenness	c=20	2.1	8.28 (8.46)	0.00565 (0.00564)
Betweenness	c=100	2.1	8.45 (8.46)	0.00552 (0.00381)
Betweenness	c=200	2.1	8.54 (8.54)	0.00796 (0.00769)
Betweenness	c=max	2.1	9.97 (9.97)	0.0113 (0.0113)
Degree	c=20	2.1	5.50 (38.8)	0.0117 (0.0103)
Degree	c=100	2.1	5.45 (35.2)	0.0166 (0.0144)
Degree	c=200	2.1	5.19 (36.2)	0.0140 (0.0123)
Degree	c=max	2.1	4.85 (35.6)	0.0170 (0.0149)
PageRank	c=20	2.1	2.27 (86.3)	0.201 (0.0751)
PageRank	c=100	2.1	1.82 (93.7)	0.425 (0.109)
PageRank	c=200	2.1	1.69 (96.1)	0.593 (0.131)
PageRank	c=max	2.1	1.50 (96.14)	0.866 (0.175)

4 Conclusion

We have proposed an agent-based evolutionary model which consists of a growing total path length under the agreement of all agents. The model assume a node of network is agent, and all agents who belongs to the network selects the new edge that newly constructed. In this model, each agent has no freedom to connect other agent, a new edge is selected by the voting. When an agent votes, it evaluates the candidate edge according to own centrality as the voting standard. There are some centrality to analysis a social network. We employ the following four population measures: closeness centrality C^C, degree centrality C^D, betweenness centrality C^B, PageRank centrality C^P.

We have simulated the evolving network based on the our model. As a result, the simulation results are different network topology for each centrality. The closeness centrality effects to the network evolution that network size becomes small. This centrality makes to select the edge connects the agents has hight centrality. That edge raises not only the connected agents' centrality, but also the other agents' centrality. That is, this centrality makes cooperation among all agents in social network, tacitly. The degree centrality has no effects cooperation among agents, since agent does not interest with the others' behavior. This is, because, the network evolutes at random, and the networks become a kind of random network. Betweenness centrality is the number of all shortest paths in the network that pass through the node. Therefore, the network topology has become that there are some simple path can pass a lot of node, and diameter of network becomes long. PageRank centrality is the principal eigenvector of the adjacency matrix of a network. Therefore, only part of the nodes are connected completely, a lot of isolated nodes exist, and it doesn't make it to the node companion with a new existing network. As a result, we have a possibility to control a network topology with our approach.

The future work is that we need to analyze the feature of these specific toplogy networks in detail, and compare with a networks in real-world.

References

1. Sen, P., Dasgupta, S., Chatterjee, A., Sreeram, P.A., Mukherjee, G., Manna, S.S.: Small-world properties of the indian railway network. Physical Review E (2003) 036106
2. Buckwalter, D.W.: Complex topology in the highway network of hungary, 1990 and 1998. Journal of Transport Geography (2001) 125–135
3. Gabora, S., Csabai, I.: The analogies of highway and computer network traffic. Physica A **307** (2002) 516–526
4. Newman, M.E.J.: The structure and function of complex networks. SIAM Review **45** (2003) 167–256
5. Barabási, A.L., Albert, R.: Emergence of scaling in random networks. Science (1999) 509–512
6. de Solla Price, D.J.: Networks of scientific papers. Science **149** (1965) 510–515
7. Freeman, L.C.: Centrality in social networks: Conceptual clarification. Social Networks **1** (1979) 215–239
8. Brin, S., Page, L.: The anatomy of a large-scale hypertextual web search engine. Computer Networks and ISDN Systems **30** (1998) 107–117
9. Haveliwala, T.H.: Efficient computation of pagerank. Stanford Technical Report (1999)
10. Wikipedia: Borda count. http://en.wikipedia.org/wiki/Borda_count (2005)

Enterprise Organization and Communication Network

Hideyuki Mizuta[1] and Fusashi Nakamura[2]

[1] IBM Tokyo Research Laboratory, 1623-14 Shimotsuruma, Yamato-shi, Kanagawa-ken 242-8502, Japan e28193@jp.ibm.com
[2] IBM Tokyo Research Laboratory, 1623-14 Shimotsuruma, Yamato-shi, Kanagawa-ken 242-8502, Japan nakamuf@jp.ibm.com

Summary. In this paper, we consider enterprise organization structures and communication networks obtained from email transaction data. Along with recent progress and popularization of Information Technology, social sciences have been experiencing great advances in methodology. It has become possible for researchers to utilize huge computational social data. However, there have been only conceptual studies in business school and very little quantitative analysis for enterprise organizations. We evaluate the strategic organization change with graph/network analysis of the communication network.

Key words: Enterprise, Organization, Communication, Network

1 Introduction

Collective human behavior causes complexity and difficulty in enterprise management. Observing an enterprise from the view point of such human connection, we notice that there are multi-layers of networks co-existing (Fig. 1). These networks are mutually related and important for business and personal activities in the enterprise.

To perform business activities strategically with strong leadership, usually there is a personnel organization in an enterprise with tree structure starting from the top which consists of board of directors and CEO, going through several level of managers, and arriving at leaves of employees. Such an organization structure can also be observed in Government and Army. In addition to the strict chain of control, a lot of modeling and managing methods of business processes such as SCM (Supply Chain Management) and CRM (Customer Relationship Management) are studied for efficient business. Similarly, organization studies are motivated to achieve business goals efficiently with strategic construction and transformation of organization considering the management policy of CEO, enterprise culture and business environments.

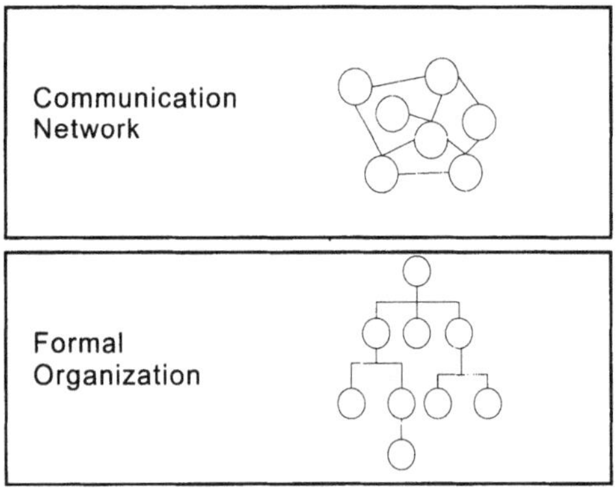

Fig. 1. Formal Organization and Communication Network

On the other hand, the value creation via flexible collaboration across organization becomes important in services business and R&D where it is difficult to apply such a modeling with typical processes. When we consider about collaboration in general, the personal connection via human-to-human communication, say Social Network, becomes a typical topic to be discussed.

Today, there is a confusion about the term "Social Network". Some people use this for tools or web sites to formulate personal communities by introducing friends to friends. In Social Science, particularly, in Computational Organization Theory, it is used to investigate communities from the view point of network. In addition, the agent-based approach has been applied to computational organization models (for example, PCANS [1] and VDT (Virtual Design Team) [2]). And also, along with the recent findings [3] on widely observed characteristics among natural or artificial networks, we often hear other expressions such as small world network [4], scale-free network, and evolving network. In this paper, we simply use the notation "communication network".

2 Communication Network

In this section, we'd like to clarify communication network which we use for the analysis. Today, there are many types of communication method. For business collaboration, we can use various level of technology from traditional face-to-face meeting and telephone to computational email, instant messaging and groupware (Fig. 2). It is still difficult to gather all these activities. But it is expected that in the near future almost all remote communication will be transferred on the IT network such as IP Phone or Web conference.

In this paper, we consider the transaction obtained from email send/receive log files together with the enterprise organization as the communication network for analysis. For the traditional social science, researchers usually investigate various communication among examinees by direct interviews or

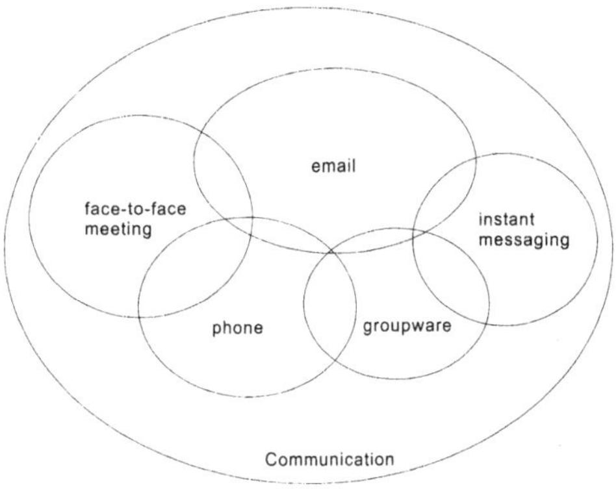

Fig. 2. Various Communication Channels

questionnaires. However such a human survey is so troublesome and time-consuming that the amount of available data is restricted. By using email transaction log files which is automatically stored on the server when people send or receive an email, we can utilize enormous communication data easily. In Econophysics, for example, researchers have revealed new characteristics of the financial market with such automated enormous data. Similarly, it is expected that huge transaction data can expand our knowledge on the communication network. In reality, the power distribution is observed both in the financial market and the Web.

We introduce a survey example of the communication network in an enterprise organization before and after strategic organization change. In our analysis, it is also important to consider the location in the strategic enterprise structure where senders or receivers of email correspond. We have analyzed huge communication from email transaction log data as described above. From a viewpoint of privacy, we transformed and aggregated the transaction data as the relationship between department for a month so that a particular employee cannot be traced. Furthermore, we did not investigate subjects and contents of email in this survey. Fig. 3 shows the data flow of personnel data and email transaction for the analysis tool in the next section.

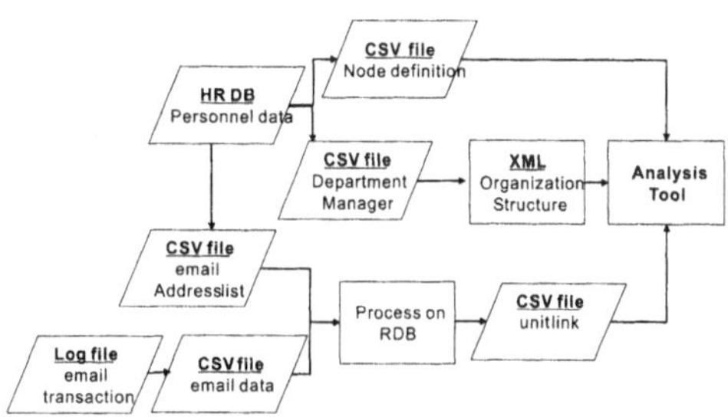

Fig. 3. Data Flow for Communication Network

3 Analysis Tools

We are developing a tool with Java to analyze the communication network over the enterprise organization. In this section, we briefly introduce a prototype of the tool (Fig. 4).

To perform the analysis in this paper, we consider simple indices for graph/network. The developed tool preprocesses input data at first. Then, it analyzes networks with implemented algorithms for network indices. Finally it visualizes the result with corresponding GUI for each index. To analyze the preprocessed data with other algorithms which have not been implemented yet, the tool can also output the preprocessed data into a file.

We have developed this tool with object oriented model so that new indices and algorithms can be added easily. In the program, each analysis class and GUI class are implemented by using fundamental interface classes.

As the preprocess, the tool aggregates email transaction data between lowest-level departments (units) described in the previous section to link information between strategic groups (nodes) which we want to investigate for consulting services. In addition, the link information which include amount of email between nodes are normalized by a number of corresponding employees in a node, or discretized with a specified threshold with the user's preferences.

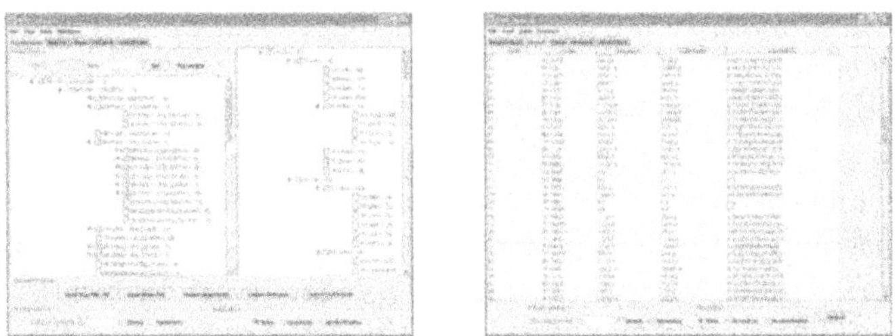

Fig. 4. COA Analyzer GUI : Organization Tree and Network Indices

Currently, this tool evaluates Input/Output Degree, Distance, and Closeness which is obtained as a inverse of the sum of distances to all nodes for each node. With an adjacent matrix $A = (a_{ij})$ where a_{ij} is the communication frequency between node n_i and node n_j and a distance matrix $D = (d_{ij})$ where d_{ij} is the geodesic distance between n_i and n_j, Degrees and Freeman's Closeness are written as follows:

$$InDegree(n_i) = \sum_j a_{ji},$$

$$OutDegree(n_i) = \sum_j a_{ij},$$

$$Closeness(n_i) = \frac{1}{\sum_j d_{ij}},$$

We also consider Power index and Faction (subgroup) with UCINET [5].

For 3D visualization of hierarchical organization and communication network, we have combined Ito's visualization tool [6] into our tool (Fig. 5).

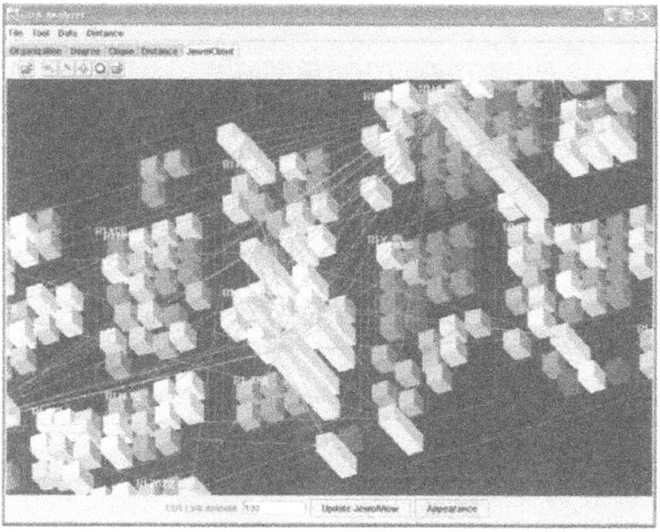

Fig. 5. Visualization

4 Example

In this section, we introduce an example of case study for organization change and communication network though we cannot reveal all details. We investigated a part of a company with consultants (IBM Business Consulting Services KK) using these data and tools described in the previous section. Especially, we directed our attention to the recent strategic organization change for strong matrix structure. We made an analysis with a node definition for this direction and observed changes appeared in the communication network.

In this survey, we used email transaction log files and a part of personnel management information to investigate organization structure for input data and to find corresponding department with each email address.

Before the organization change, the matrix structure was not balanced and one direction (product axis) was dominant, which caused customer facing various separated products without a unified service. The focused organization change is intended to strengthen another direction (sector axis) to balance the matrix structure and to provide one face services to customer in each sector.

Hence, we began the analysis by investigating changes in Degrees of the communication network along these two directions. Summarized and scaled results are shown in Fig. 6.

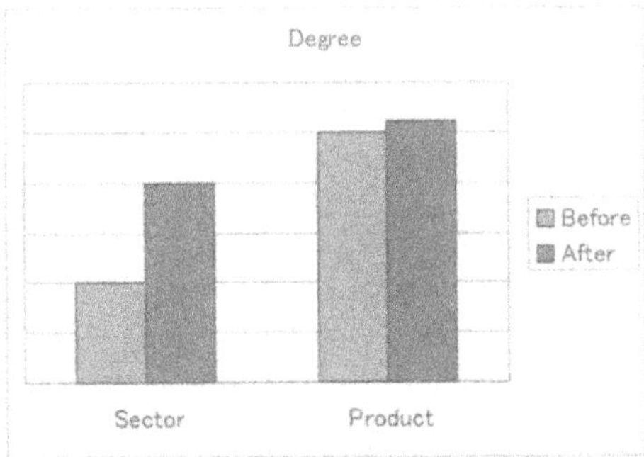

Fig. 6. Degree Change in Product and Sector Direction

We can observe a significant increase of the communication along sector from the graph. On the other hand, this change does not cause a undesirable side effect on the communication along product.

Next, we analyzed changes in Total Closeness (Fig. 7).

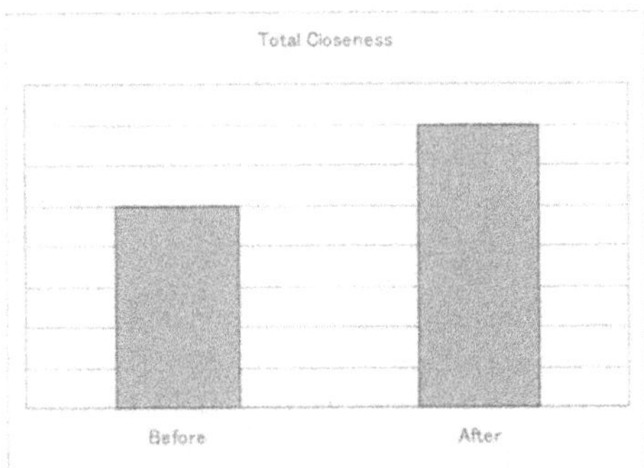

Fig. 7. Total Closeness Change

Together with the previous results which indicate both communication in the matrix organization with sector and product axes become strong and balanced, we can consider that the whole organization network becomes more condensed one.

Finally, we show a number of nodes in subgroups divided by UCINET as Faction for each sector (Fig. 8). It can be considered that nodes in a same subgroup communicate frequently. Before the organization change, a subgroup

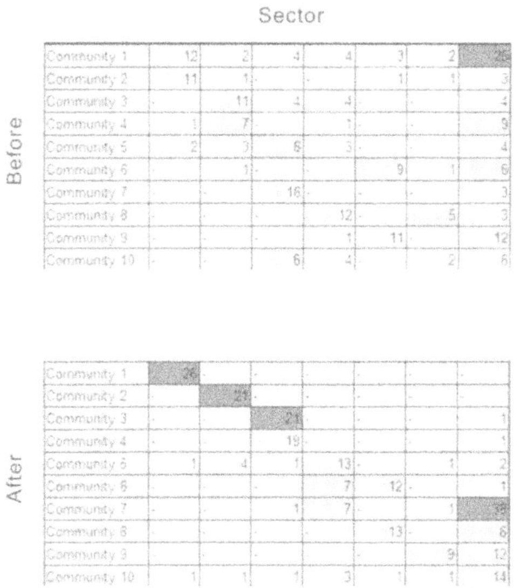

Fig. 8. Subgroup Change

contains nodes of various sectors. However, after the organization change, a subgroup contains nodes of one sector or a few sectors. This result agrees with the degrees along sector increase. Furthermore, it shows that increased communication is strongly correlated with strategically determined sectors.

5 Concluding Remarks

In this survey, we observed the effects of an organization change by the top-down strategic decision over the communication network with simple indices.

Traditional studies on enterprise organization such as matrix structure are restricted within descriptive intuition usually performed as case studies at Business School, and lack quantitative analysis.

With our survey though only with simple indices at now, effects of the organization change can be analyzed by using enormous email transaction data which can be collected automatically. More business-related analysis with complex indices remains as future works.

In addition, the distribution of Degree of the communication network surveyed in this paper are shown in Fig. 9 sorted by order of degree from the largest node to 100th node using a log-log scale.

This result indicates a power law distribution and that the communication network over enterprise organization also has a characteristic of scale-free as many other natural networks though this network is constructed on a strategically determined organization structure. Future analysis on such characteris-

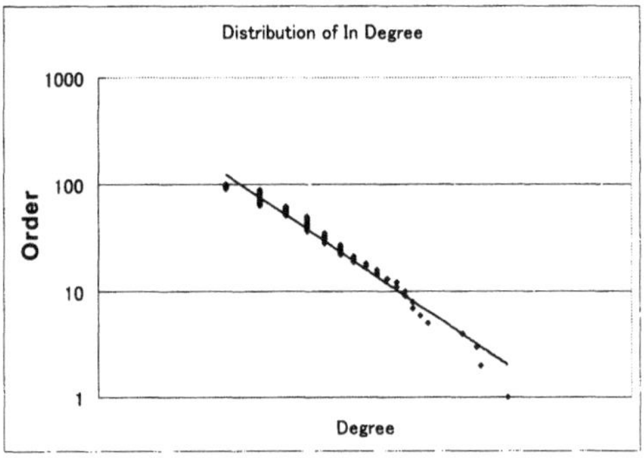

Fig. 9. Power Law in Degree Distribution

tics of communication network itself and the influence upon business activities is important.

References

1. Carley K, Krackhardt D (1998) A PCANS Model of Structure in Organizations. In: Proceedings of the 1998 International Symposium on Command and Control Research and Technology. June, Monterey, CA
2. Levitt RE (2004) Computational Modeling of Organizations Comes of Age. Computational & Mathematical Organization Theory 10:127–145
3. Barabasi AL (2002) Linked, The New Science of Networks. Perseus, Cambridge
4. Watts DJ (1999) Small Worlds - The Dynamics of Networks between Order and Randomness. Princeton University Press, Princeton, NJ
5. UCINET (Analytic Technologies), http://www.analytictech.com
6. Itoh T, Yamaguchi Y, Ikehata Y, Kajinaga Y (2004) Hierarchical Data Visualization Using a Fast Rectangle-Packing Algorithm. IEEE Transactions on Visualization and Computer Graphics 10:302–313

Scaling properties of Two-sided Matching with Compromise

Tomoko Fuku, Akira Namatame

{g43036, nama}@nda.ac.jp
Department of Computer Science, National Defense Academy, Yokosuka, 239-8686, Japan

Abstract:
In this paper, we propose a new algorithm for two-sided matching problems with some compromise. It is an important issue such as how far agents seek their own interest in a competitive environment? There are overwhelming evidences that support peoples are also motivated by concerns for fairness and reciprocity. We showed that compromise, an individually irrational behavior, improves the welfare of others. We also obtain large-scale properties of some two-sided matching algorithms. We show some compromises of individuals increase global welfare. The optimal compromise level is designed so that the social utility is maximized.

1. Introduction

Some researchers have started to take a direct role in issues of designing market, e.g. labor market, a venue for bilateral trading require a proper matching. But, in considering the design of markets, is extremely complex. Markets evolve, but they are also designed. The complexity of market design comes from many factors, especially strategic behaviors of participants. A market is two-sided if there are two sets of agents, and if an agent from one side of the market can be matched only with an agent from the other side. One of the main functions of many markets is to match one kind of agent with another: e.g. students and colleges, workers and firms, marriageable men and women. A two-sided matching model is introduced by Gale and Shapley(1962), and they invented the deferred acceptance algorithm.

Much of two-sided matching theory is concerned with determining the conditions under which stable matching exists, and with what algorithms these matching can be achieved. The relationship between the concept of Pareto optimality and the stability of a matching has been also investigated. Pareto optimality requires that no change exists that better every individual in the population. The concept of a stable matching is stronger than that of a Pareto optimal matching, in that every stable matching is Pareto optimal, but not every Pareto optimal matching is stable. Pareto optimality requires that no two individuals wish to elope together and would receive the consent of their partners. Stable matching, by contrast, requires that no two individuals wish to elope together, whether or not their partners would consent.

We also obtain large-scale properties of the proposed algorithm. It is an important issue how far agents seek their own interests in a competitive environment. There are overwhelming evidences that support peoples are also motivated by concerns for fairness and reciprocity. The reasonable compromise level is obtained as the function of the size of the group so that the social utility should be maximized.

2. Deferred Acceptance Algorithm and Its Properties

There are two disjoint sets of agents, groups of $M = \{m_1,..., m_n\}$, and $W = \{w_1,..., w_n\}$. Associated with each side is the number of positions they have to offer. Agents on each side of the market have (transitive and strict) preferences over agents on the other side. Gale and Shapley derived a clever algorithm (described below) designed to create efficient pairings. Their algorithm matches students and schools in such a way that no student wishes to leave her current school for an alternative institution that would be willing to admit her. Subsequent authors expanded upon Gale and Shapley's work, extending their theoretical framework while applying two-sided matching theory to problems ranging from labor markets to human courtship and marriage.

A pair m and w are said to *block* the matching μ if they are not matched by μ, but prefer one another to their mates as assigned by matching μ. Put another way, given *matching* μ, a blocking pair is a pair that would willingly abandon their mates as determined by μ and elope instead with one another. Finally, the matching μ is defined as ***stable*** if it is not blocked by any pair of agents.

We outline the ***Deferred Acceptance Algorithm*** (***DA***). The matching procedure of DA proceeds repeatedly through the following steps.

<Step 1> Each man not currently engaged displays to his favorite woman that has not already rejected him. If no acceptable women remain, he remains unmated.

<Step 2> Each woman who has received one or more courtship displays in this round rejects all but her highest-ranked acceptable man. This may involve rejecting a previously engaged man.

<Step n> After a large number of rounds, no new displays will be made.

At some point, the algorithm terminates. All women are paired with men to whom they are currently engaged; individuals not engaged remain unmated.

The matching μ generated in this way is easily seen to be stable. No men wishes to leave his mate at μ for a female who prefers him to her mate at μ, because each male reached his current mate by sequentially courting females in order of preference. No women wishes to leave her mate at μ for a man who prefers her to his mate at μ, because she will have already received a courtship display from any men who is not matched to a woman that he prefers to her.

Reversing the algorithm, so that the women display and the men accept or reject courtships, will also lead to a stable matching; this matching may be a different one than that found by the man-courtship form of the algorithm. However, the set of individuals remaining unmated is the same in every stable matching of any given monogamous mating system.

As above-mentioned, Deferred Acceptance Algorithm produces either **(1) Man-optimal stable matching,** or **(2) Woman-optimal stable matching**. That is (1) the matching h_m produced by the deferred acceptance algorithm with man proposing is the M-optimal stable matching, or, (2) the W-optimal stable matching is the matching h_w produced when the woman propose (Gale and Shapley(1962)). And it's emerged that the best outcome for one side of the market is the worst for the other, i.e. M-optimal stable matching is the worst for women. And W-optimal stable matching is the worst for men (Knuth (1972)).

3. Two-sided Matching with Compromise

In this section, we propose two new algorithms based on the hypothesis bounded rationality.

<Definition of Utility>

Before we state our algorithms, we define utility measures.

N : the size of each group

T : the preference level of the partner to be matched ($1 \leq T \leq N$)

(1) Individual utility (agent i)

$$U_i = 1 - T_i / N$$

(2) Utility of a pair of agent i and agent j

$$P_{ij} = U_i \times U_j$$

(3) Social utility

$$SU = \sum_{ij=1}^{N} P_{ij} \tag{3.1}$$

3.1 Matching with Compromise : MAC

It's natural that the optimal strategy is to behave as if the individuals can profit the maximum. But like prisoner's dilemma game everyone knows, the social best optimum is to act in concert. Similarly, in the marriage problem we wonder if the social optimal point exists when the individuals compromise acceptable level. It happens that no compromise result in the lowest social utility because of the large difference satisfaction with disappointment, and increasing matching paired slowly. Moreover, it is sure that more compromise result the lowest social utility because many individuals can not be satisfied the outcome. However we can compromise at individual level properly, most of individuals somewhat can increase their satisfaction.

Necessarily, we suppose that we can get the most social utility by increasing the pairs with somewhat satisfaction. For the present, it delays the discussion of this algorithm's efficacy till later, we thought about the value of the most optimal compromising level. It follows the simulation result.

A proposed woman accepts the proposal within the acceptable level, otherwise she rejects it (holding which is allowed in DA is not permitted).

In order to obtain the reasonable level of compromise, we simulated by changing the compromise level gradually and obtained the social utility as shown in Figure 1. We obtain the optimal level of compromise C by changing the group size up to N=1000 as shown.

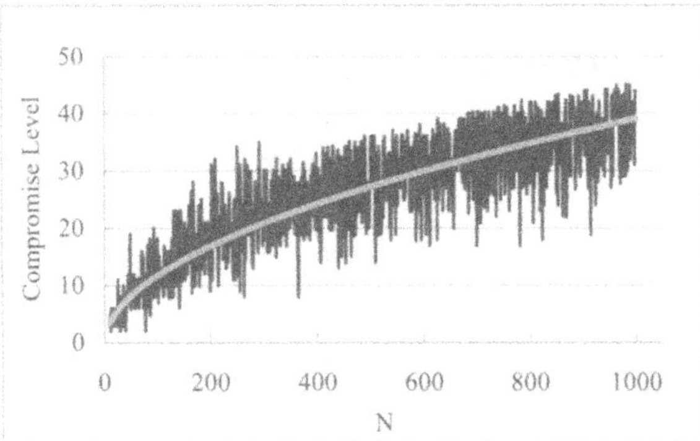

Figure 1. The optimal compromising level

Although there are some fluctuations for the optimal compromise level, it is roughly approximated as the function of the group size N as follows;

The relation between compromise (C) and N is approximation by

$$C = 1.077 N^{0.5197} \cong \alpha\sqrt{N} \ (\alpha \cong 1.2) \tag{3.2}$$

3.2 Advanced Deferred Acceptance : ADA

We also propose a new algorithm by modifying DA (Deferred Acceptance). In DA, woman receive man's propose which is the woman's list even if his grade is low. Then, if woman can be proposed the man who is higher-rank than engaged man in his list, she is mated with new man. All agents can be mated, then, they have to marry with engaged partner. This feature is that woman can wait higher-rank partner by the time limit, but that woman must marry with unpleasant man if she can't meet higher-rank man. This is like the double-edged sword. So, we set the borderline, we tried relieving the woman's disaffection. We suppose we can progress the social utility. Each individual usually have the minimum standard, but we set it on purpose. A proposed woman accepts the proposal within the acceptable level, otherwise she rejects it.

4. Simulation Results

4.1 Uncorrelated preferences: Randomly Generated Preferences

We generate preference orders of all agents randomly. Agents are matched using three algorithm simulation (*Deferred Acceptance* (*DA*), *Matching with Compromise* (*MAC*), *Advanced Deferred Acceptance* (*ADA*). We compare the utility distribution of all agents (both man and woman), and the social utility. It follows the simulation result (N = 1,000), that is the distribution of availability.

We show simulation result in Figure 2, 3, 4, when man propose woman. Figure 2 shows the utility distribution of man. Since the DA produced the M-optimal, most men receive the utility that closes to 1. It is the worst for woman. On the other hand, MAC and ADA produced the balanced matching, and the utilities of both man and woman are high.

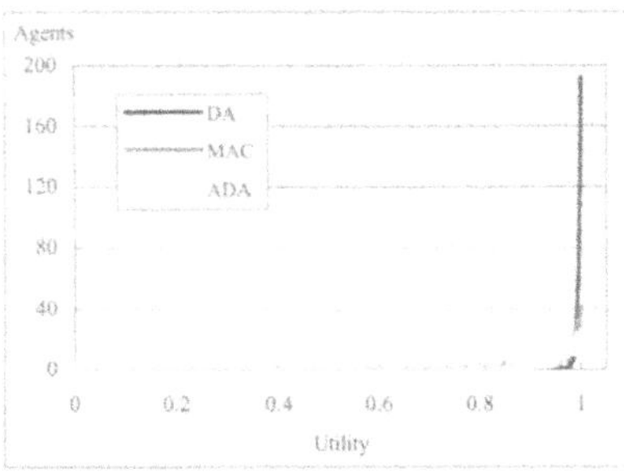

Figure 2. The utility distribution(men)

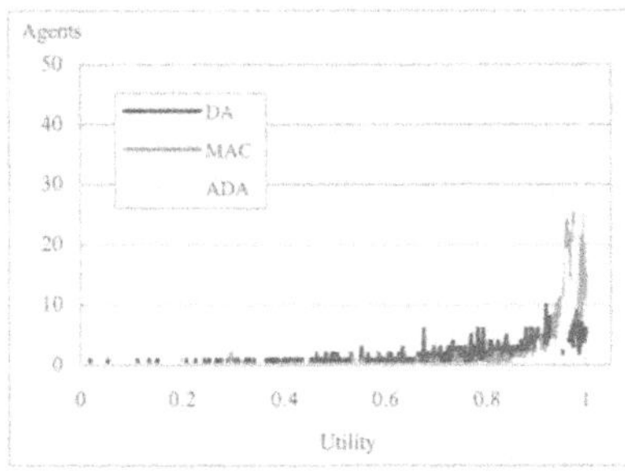

Figure 3. The utility distribution(women)

It is natural that many proposing agents can satisfy the outcome, but proposed agents can not. In other word, DA have many satisfied proposing agents, few proposed agents. Consequently, the utility of the Pair hold lower than MAC and ADA. Then, we can get the higher utility of pair with some compromises or borderlines.

Figure 4 shows social utilities of two algorithms are higher than in DA. Especially, the more the circle scale is bigger. When the circle scale is small, in MAC social utility can be lower, because compromise level is large percentage against N. But we can get high social utility in MAC in large circle.

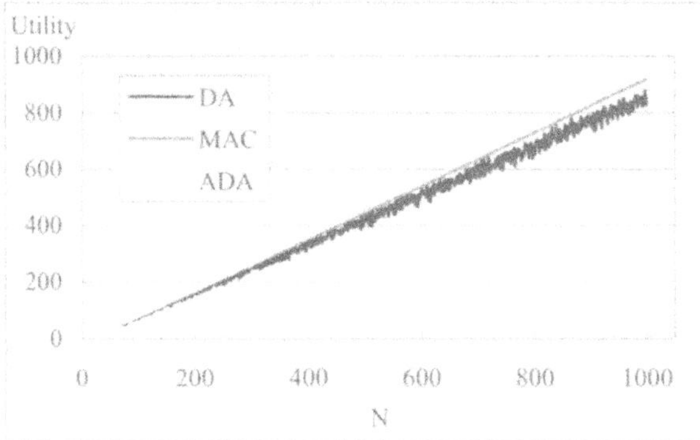

Figure 4. Comparison of social utility of three algorithms

4.2 Correlated preferences

In the foregoing paragraph, the preferences are generated at random. There is not so much correlation in the preference orders between each agent, and the possibility that they collide is few for that. Therefore, the match with high social effect was achieved. However, there are often actually some correlations between related to agent's preference. For instance, the case where popularity concentrates on some certain people is not few. In this section, there is bias in the preference relationship of the agent, and the case where all agents have a high preference relationship for a specific agent of the other party group is taken up, and the relation between diversity related to the performance of the algorithm is clarified. Figure below is a comparison of social utility when popularity concentrated on the high rank 10% and 33% (θ=0.1, 0.33).

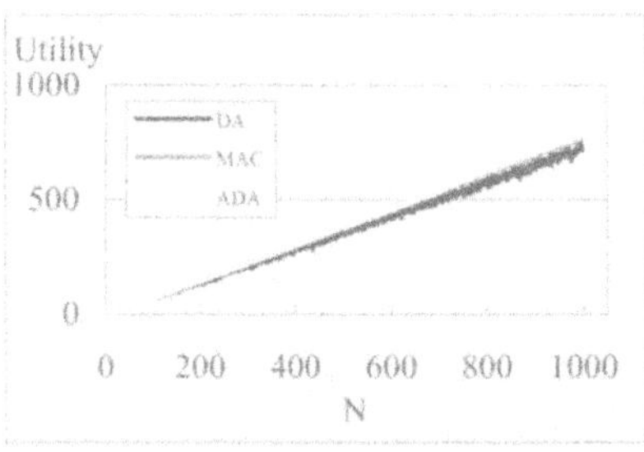

Figure 5. Comparison of social utility ($\theta = 0.1$)

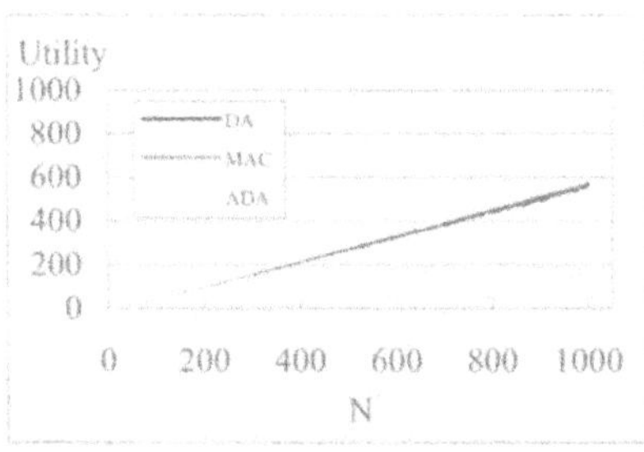

Figure 6. Comparison of social utility ($\theta = 0.33$)

The match with high social effect is possible when the value of θ that shows a popular concentrated level to a certain specific person is small. When the value of

θ grows on the other hand, a social effect falls. It is understood that a preferable match of a completely different group that there is not so much correlation in the Eras relation between one person one is more possible than that of a homogeneous group that the Eras relations between individuals look like from this. In a word, the effect with a high match (match based on the compromise) algorithm based on a limited rationality one person one is demonstrated in the society where the preference relationship is completely different. However, when the bias relation between people is similar and a lot of bias relations of the person are the same, the difference on the performance of the match algorithm becomes small.

5. Theoretical Considerations of the Optimal Compromise Level

In section 3, we obtained the relationship between the optimal compromise level and the population size by simulation which obtains in (3.2). In this section, we obtain it theoretical analysis. We define the following terminologies:

$g(t)$: The number of paired couples at the t step,

$f(t)$: The number of unmatched agents at the t step

C : The compromise level

In the case, the contents of preference lists each agents have are at random, paired couple arise at random, too. Each side scale is N agents, the compromise level is C, we suppose. g is the number of paired couples on the step, f describes the number of other free agents.

<Step1> Each N agents on one side propose the other side best partner, then, the probability of accepting proposal is C/N. So, on the first step, the number of paired-couple is

$$g(0) = N \times C / N = C\,, \tag{5.1}$$

and the number of free agents (who are still single, non-couple) is

$$f(0) = N - g(0) = N - C\,. \tag{5.2}$$

<Step t> The agents of $f(t-1)$ propose the tth partners of each preference lists. The probability that the tth partners of each preference lists is still free is $f(t-1)/N$, and the probability of the accepting proposal is $(C{+}t)/N$.

Then, the number of paired couple on the step t is

$$g(t) = f(t-1) \times \frac{C+t}{N} \times \frac{f(t-1)}{N} \tag{5.3}$$

On this step, the number of free agents is

$$f(t) = f(t-1) - g(t)$$

That is

$$f(t) = f(t-1) \times \{1 - f(t-1) \times \frac{(C+t)}{N^2}\} \tag{5.4}$$

If we solve the recursive equation (5.4), we have $C \cong 1.3\sqrt{N}$. Therefore we could get the nice approximation in (3.2).

6. Conclusion

Deferred Acceptance has universally used since it suggested for 40 years. To be sure, this is very brilliant method for two sided matching, but it has a demerit which is a *M-optimal* algorithm. This research pursued that we produced the moderate equality for each side. We suggested two algorithms. The first is the matching of the compromise MAC, the second is the matching of using borderline in DA, ADA. We use agent-based simulation, we inspect each algorithm's quality and social utility. We could show *pair-optimal* matching by using MAC. I mean, to be sure compromise behavior is an individually irrational, but is not in macro world. Compromise behavior link collaboration, and eventuate high efficiency. The main reason is sure that much compromise profit better partner not the best partner. For many agents, the best partner exist only one, but better partner exist more. Many agents can moderately satisfy, therefore, social utility is getting better globally.

Besides, we produced ADA as an improved version, we injected the borderline of the proposed side, could overcome the disadvantage the side of the proposed. As a result, we could produce the welfare of the society. In addition, the balance of proposing side and proposed side can keep well. It is sure that proposing side utility gets down totally, but commensurately proposed side utility is getting better. Globally, we can get higher social utility outcome. It may natural that compromise or borderline profit social utility globally, nevertheless DA is the best individually. But we will not insist the new algorithms performance.

References

1. CrawfordV.P. and Knoer E.M (1981) *Job matching with heterogeneous firms and workers*. Econometrica 49: 437–450
2. Marie-jose Omero, Michael Dzierzawa, Matteo Marsili, Yi-Cheng Zhang (2004) *Scaling Behavior in the Stable Marriage Problem*. cond-mat/9708181
3. Roth A.E. (1982) *The economics of matching: Stability and incentives*. Mathematics Operations Research 7: 617–628
4. Roth A.E. (1989) *Two-sided matching with incomplete information about others' preferences*, Games Economic Behavior 1: 191–209
5. Roth A.E. and Sotomayor (1990) *Two-sided Matching: A Study in Game-theoretic Modeling and Analysis*. Econometric Society Monographs No. 18. New York: Cambridge University Press
6. Shapley L.S. and Shubik M. (1972) *The assignment game I: The core*. Int. J. Game Theory 1: 111–130

Learning Agents

Reinforcement Learning on Monopolistic Intermediary Games: Subject Experiments and Simulation*

Atsushi Iwasaki[1], Kazuhito Ogawa[2], Makoto Yokoo[1], and Sobei H. Oda[3]

[1] Kyushu University
[2] Hiroshima City University
[3] Kyoto Sangyo University

1.1 Introduction

Decision making processes of consumers and firms have emerged as a prominent field in multi-agent systems. This approach relaxes the strict assumptions in traditional economic theory. However, little attention has been paid to the validation of the simulation itself, because of the difficulty of acquiring field data which is compared with simulated markets.

We provide comparable data with the observed actions of subjects and validate an individual decision-making process. Recently, a variety of learning models have been proposed and examined by some economists. However, they only have examined simple games such as two-person games with few alternatives, all of which have a unique equilibrium such as the Prisoners' Dilemma (Roth and Erev [8]). We utilize more complicated games wherein in a player acts as a monopolistic intermediary against given but unknown supply and demand functions and then chooses bid and ask prices repeatedly, Furthermore we investigate the descriptive power and limitations of the simple reinforcement learning (REL) model proposed by Erev and Roth [3].

1.2 Theory, Experiments, and Results

The market equilibrium is determined at the intersection of the supply and demand curves and a Walrasian auctioneer is assumed and expected to re-

* This research was supported by the Open Research Centre "Experimental Economics: A new method of teaching economics and the research on its impact on society, " Kyoto Sangyo University, JSPS, Grant-in-Aid for Scientific Research (B), 13480115, JSPS Grant-in-Aid for Doctoral Candidate Fellows, FY2002-FY2004, and Foundation for the Fusion of Science and Technology (http://www.fost.or.jp/).

peat auctions until he discovers the equilibrium price. This is true only if the auctioneer is a sincere altruist and to maximize the total surplus.

Instead, Spulber [10] assumes an intermediary who organizes the market in order to earn profits. Suppose there is a monopolistic intermediary *in* the market. If she is aware of the supply and demand functions, she can readily calculate the bid price w (the price at which she buys the commodity from its sellers) and the ask price p (the price at which she sells the commodity to its buyers) that maximize her profit margin. The Walrasian equilibrium is realized only if she chooses the intersection of the supply and demand functions.

Ogawa et al. [6] have designed a single-person game in which a player, acting as the monopolistic intermediary of a commodity, determines w and p to earn a profit $\pi = py - wx$, where x represents the number of units of the commodity bought from the producers and while y denotes for the number of units of the commodity sold to the consumers.

They developed two treatments: the *stock market* treatment and the *fish market* treatment. In the former, subjects are allowed to refuse to buy or sell some units at the prices they have chosen, whereas in the latter, they must accept all sell orders at the bid price they determine; $\pi(w,p) = (p-w)\min\{S(w), D(p)\}$ under the stock market treatment, and $\pi(w,p) = p\min\{S(w), D(p)\} - wS(w)$ under the fish market treatment. Here, $S(w)$ and $D(p)$ represent the market supply function and the market demand function, respectively. The only difference between the two treatments is that the dead inventory, or the number of units that are bought but not sold, given by $\min\{0, S(w) - D(p)\}$, is always zero under the stock market treatment, whereas it can be positive under the fish market treatment.

Under either market treatment, $S(w) = D(p)$ is a necessary condition for maximizing the profit $\pi(w,p)$. The only difference is that for any price pair (w,p), the profit either is lower under the fish market treatment or is identical for either treatment. Arriving at a shortfall of profit due to $S(w) \neq D(p)$ is unconditionally more difficult under the fish market treatment.

Ogawa et al. [6] introduced the stepped supply and demand functions (Fig. 1.1 (a)). Stepped functions make experiments tractable, while yielding some locally optimal pairs of bid and ask prices. Since this affects our experimental and simulation results, we discuss both a globally and locally optimal points.

The supply and demand functions used by Ogawa et al. [6] were common to both market treatments. In each session, subjects were asked to choose two integers between 1 and 100 as w and p. Since the system was programmed to reject inputs where $w > p$ and request another input, there were 4,900 pairs of prices to choose from.

Subjects participated in the stock market or fish market treatments. Each subject played the monopolistic intermediary for 20 rounds under each treatment. Details of the experiment were provided in an instruction manual and read out by the instructor prior to the subjects playing 10 training rounds. Before the experiment, subjects were only informed that $S(w)$ was non-decreasing and $D(p)$ was non-increasing. Further, to make each subject's

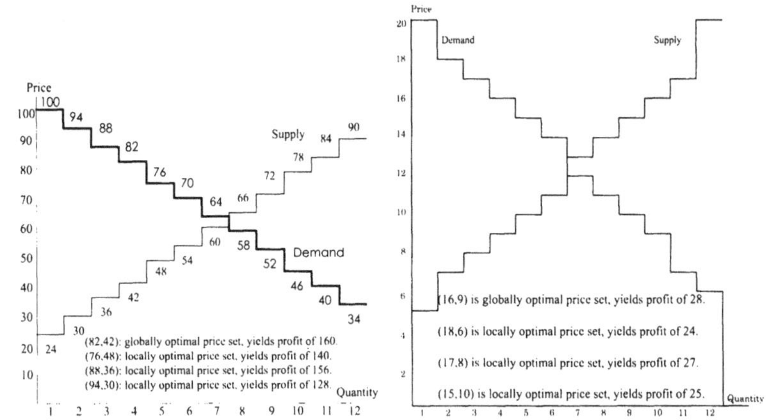

(a) Experiment. Source: Ogawa et al. [6]

(b) Computer simulation.

Fig. 1.1. Functions used under experiments and under computer simulation.

decision-making easier, the history of his or her choices and the corresponding results (w, p, x, y and π) were displayed on their monitors for each session.

Ogawa et al. [6] found following experimental results. Typically, a subject changed prices more frequently in the first 10 rounds than in the subsequent 10 rounds. Finally she arrived at the prices that satisfied the supply and demand. In order to find such price sets, most subjects utilized *"two-stage searching"*; they search for the bid and ask prices that equalize the purchasing and selling quantities. Maintaining the trading quantity constant, they then adjust the prices to realize a locally or globally maximum profit.

1.3 Decision Making Models for Computer Simulation

To compare the experimental results with our simulation results, simulated players offer integers between 1 and 20 as bid and ask prices (Fig 1.1 (b)). Since the constraint of human experiments applied to computer simulation, simulated agents cannot offer price pairs such that $w < p$. In this case, the simulated player encounters 190 possible price pairs to choose from.

It is, of course, possible to constract a simulation setting identical to the experimental setting. However, when an experiment has a very large number of alternatives, it is often preferable to aggregate the experimental data and simplify the actual strategy space because the raw data is simplified only before we can analyze it at a confidence level (Grosskopf [1], Roth and Erev [8]).

We now introduce the model based on the REL proposed by Erev and Roth [3] and its variations. In this model, simulated agents choose their alternatives stochastically according to their propensities, which represent the alternatives that enable them to receive higher payoffs. Agents choose their strategies through trial and error in order to obtain higher payoffs, thereby learning to play better strategies more often and worse strategies less often.

In particular, a simulated player independently chooses bid and ask prices from integers between 1 and 20 in each round. She decides both the prices depending on her propensities and posts them to buyers and sellers. For simplicity, we explain the basis of selecting an ask price.

When a simulated agent i chooses a price k in round t, she receives a payoff given by $\rho(t)$. Given her payoff, her propensity $q_{ij}(t)$ is updated according to

$$q_{ij}(t+1) = \begin{cases} q_{ij}(t)[1-\gamma(t)] + \rho(t)\gamma(t) & \text{if } j = k, \\ q_{ij}(t)[1-\gamma(t)] & \text{otherwise,} \end{cases} \tag{1.1}$$

where $\gamma(t) = \frac{C_{ij}(t)+\eta/m_i}{C_{ij}(t)+\eta/m_i+1}$. Here, η is an initial strength parameter that represents the weight of the initial propensities and m_i is the number of strategies, which is equal to 20 in the games, and $C_{ij}(t)$ represents the number of times agent i chose an ask price of j up to round t.

After the simulated player chooses the current bid and ask prices, the current supply x, demand y, trading volume z, and profit π are determined according to the simplified supply and demand curves and profit functions.

The probability with which agent i chooses price $k \in [1, 20]$ in round t is

$$p_{ik}(t) = e^{q_{ik}(t)\frac{\lambda}{S_n(t)}} / \sum e^{q_{ij}(t)\frac{\lambda}{S_n(t)}}, \tag{1.2}$$

where λ represents reinforcement sensitivity. $S_n(t)$ is calculated as the average absolute difference between the recent payoff and the accumulated average payoff in the first t rounds, and n represents the number of possible outcomes. $S_n(t+1) = S_n(t)\gamma'(t) + |A_n(t) - \rho(t)|\,(1-\gamma'(t))$, where $\gamma'(t) = \frac{t+\eta}{t+\eta+1}$. (λ, η) are fixed at (4.5, 200), which are the best parameters from the examination of several decision-making problems by Erev and Barron [2]. The accumulated average payoff in the first t rounds is calculated as $A_n(t+1) = \rho(t)\gamma'(t) + (1-\gamma'(t))A_n(t)$.

The initial propensities are given by $q_{ik}(1) = A_n(1)$, where $A_n(1)$ represents the expected payoff, given a random choice. Under the stock market and fish market treatments, $A_n(1)$ were 7.04 and -12.84, respectively. Therefore, each agent has initial propensities with which she chooses each strategy with an equal probability. $S_n(1)$ is given by $= \sum_{j=1}^{190} |\sigma_j - A_j(1)|/190$, where 190 represents the number of potential bid and ask price pairs, subject to $p > w$, $p \in [1, 20]$, and $w \in [1, 20]$. σ_j indicates the profit of the jth pair of bid and ask prices. $S_n(1)$ under the stock market and the fish market were 6.33 and 28.67, respectively.

We term this basic reinforcement learning model "REL−1". On the other hand, the values of the two prices selected by human subjects were neither too high nor too low. Therefore, we incorporate this observation and define the model as "REL−2". In this model, simulated players can choose $p \in [6, 18]$ and $w \in [3, 13]$. The intervals comes from the experimental result.

In addition, we incorporate a deterministic rule into the REL. This model partially captures the *two-stage searching* strategy. In this model, a simulated player chooses bid and ask prices according to the REL in the first 5 rounds. In the remaining rounds, she chooses these prices taking into consideration the previous supply and demand; If $S(w_{t-1}) > D(p_{t-1})$, she offers w_t in accordance with the REL scheme, while she offers an ask price of $p_t = p_{t-1}$. If the prices chosen are such that $w_t \geq p_t$, we set $p_t = w_t + 1$. If $S(w_{t-1}) < D(p_{t-1})$, she offers p_t in accordance with the REL scheme, while she sets a bid price of $w_t = w_{t-1}$. If the prices chosen are such that $w_t \geq p_t$, we set $w_t = p_t - 1$. If $S(w_{t-1}) = D(p_{t-1})$, she does not change the ask price but lowers the bid price to $w_t = w_{t-1} - 1$ from the 6th to the 10th round. Then, she does not change either prices after the 11th round. Lowering the bid price decreases the dead inventory loss. This is very effective in reproducing the subjects' behavior, especially under the fish market treatment.

Furthermore, similar to the case of the REL, we define "D-REL-1" and D-REL-2. In the "D-REL-2," the number of possible bid and ask price pairs are limited as in the case of "REL−2".

The introduction of the $S(w) = D(p)$ rule is natural. Regardless of the importance of the inventory, most market experiments ignored it. Similarly, studies that compare the behavior of human subjects and simulated players rarely dealt with the price formation of firms. It is an open question as to how to deal with the inventory. Since holding inventory is an opportunity cost, it is quite common to reduce the inventory as possible as real firms can. Just In Time system is a good example. Furthermore, other experimental research (Ogawa et al. [7]) also report that subjects apply this rule intentionally.

We now explain the Best Response (BR) strategy; when an agent chooses a bid price, the agent is assumed to take the ask price that yields the highest profit, given the bid price. The game has one globally optimal profit and some locally optimal profits as found in Fig 1.1. Thus, we introduce two BR predictions to benchmark the REL model,[4] BR-1 and BR-2. In the BR-1 the average profit of (locally and globally) optimal price sets are chosen. On the other hand, the BR-2 always chooses the globally optimal price set.

1.4 Estimating Simulation Results with Mean Squared Deviation

In order to statistically assess the accuracy of the model, we use the mean squared deviation technique (MSD, Selten [9]). We introduce the definition of MSD_{price} and MSD_{profit} from Feltovich [4]. Assuming that $P(p = \beta | w = \alpha)$

[4] Iwasaki et al. [5] suggest that a unique equilibrium prediction is more likely to fail than the REL prediction.

Stock	MSD_{price}	MSD_{profit}	Fish	MSD_{price}	MSD_{profit}
REL-1	0.1817	0.0943	REL-1	0.1434	0.5733
REL-2	0.1256	0.0393	REL-2	0.0677	0.1206
BR-1	0.3011	0.6471	BR-1	0.3238	3.7267
BR-2	0.1537	**0.0144**	BR-2	0.0921	**0.0572**
D-REL-1	0.1609	0.0829	D-REL-1	0.1276	0.4169
D-REL-2	**0.1041**	0.0295	D-REL-2	**0.0545**	0.0746

Table 1.1. The MSD results. Hereafter the boldface type indicates the lowest score.

stock	first 10 rounds	last 10 rounds	fish	first 10 rounds	last 10 rounds
REL-1	0.1996	0.2338	REL-1	0.1484	0.1599
REL-2	0.1324	**0.1724**	REL-2	**0.0770**	0.0740
BR-1	0.3131	0.3205	BR-1	0.3654	0.2992
BR-2	0.1618	0.1853	BR-2	0.0997	0.0955
D-REL-1	0.1905	0.2441	D-REL-1	0.1537	0.1527
D-REL-2	**0.1233**	0.1769	D-REL-2	0.0772	**0.0701**

Table 1.2. The MSD_{price} scores in first and last 10 rounds.

is the conditional probability, an agent chooses a bid price of α, she chooses an ask price of β. MSD_{price} score is defined as $\frac{1}{19*2}\sum_{\alpha=1}^{19}\sum_{\beta>\alpha}^{\beta=20}(P_{pred}(p = \beta|w = \alpha) - P_{obs}(p = \beta|w = \alpha))^2$, where $P_{obs}(\cdot)$ denotes an actually observed probability in subject experiments, while $P_{pred}(\cdot)$ denotes a simulated probability by using either the REL, or BR.

MSD_{profit} is defined as $\frac{1}{20}\sum_{i=1}^{20}\left(\frac{profit_{obs,i}-profit_{pred,i}}{\max profit}\right)^2$,where $profit_{obs,i}$ denotes an actually observed profit in subject experiments, $profit_{pred,i}$ denotes a simulated profit from the REL, or BR, $\max profit$ represents the theoretically maximum profit, namely, 28, and i denotes the number of rounds.

Both the scores have an interval of $[0, 1]$ and the closer they are to zero, the better the model describes the subjects' behavior. The simulation model that reproduces the behavior of human subjects well lowers both the scores.

To calculate MSD of REL-1, REL-2, D-REL-1, and D-REL-2, we employ the following steps: first, 500 simulated players choose bid and ask prices in 20 rounds. Using this data, we calculate MSD_{price} and MSD_{profit}. We then change the seed value and repeat these steps 50 times.

Table 1.1 shows that under both treatments, D-REL-2 and BR-2 yield the lowest MSD_{price} and MSD_{profit} values, respectively. The result of MSD_{price} comes from the fact that "D-REL-2" has larger search space than "BR-2". For MSD_{profit}, the outcome arises the fact that in the later rounds, more subjects chose the globally optimal price set.

Table 1.2 shows that MSD_{price} in the first and second halves. Under the stock market treatment, the lowest MSD_{price} is given by D-REL-2 in the first half and REL-2 in the second half. Under the fish market treatment the

stock	first 10 rounds	last 10 rounds	fish	first 10 rounds	last 10 rounds
REL-1	0.1471	0.0414	REL-1	0.8096	0.3369
REL-2	0.0505	0.0259	REL-2	0.1624	0.0789
BR-1	0.5103	0.7839	BR-1	3.6164	3.8371
BR-2	**0.0231**	**0.0057**	BR-2	**0.0694**	0.0450
D-REL-1	0.1485	0.0474	D-REL-1	0.7433	0.0905
D-REL-2	0.0530	0.0060	D-REL-2	0.1419	**0.0074**

Table 1.3. The MSD_{profit} scores in first and last 10 rounds.

lowest MSD_{price} is given by REL-2 in the first half and D-REL-2 in the second half, respectively. Table 1.3 shows that under the stock market treatment, the MSD_{profit} of BR-2 is the lowest in the first and second halves. Further, under the fish market treatment, the MSD_{profit} scores of BR-2 and D-REL-2 were the lowest in the first and second halves, respectively.

Finally, we explain the ratio of $S\ (w) = D\ (p)$ that the simulated players chose and show that the REL does not accurately reproduce *two-stage searching* strategy nor arrive at a locally or globally optimal price set. While human subjects realized $S\ (w) = D\ (p)$ more frequently in the latter 10 rounds than they did in the first 10 rounds, the simulated players with the REL-1 and REL-2 did not realize $S\ (w) = D\ (p)$. Even in the last 5 rounds, under the stock market treatment, $S\ (w) = D\ (p)$ was realized at most twice, whereas human subjects realized $S\ (w) = D\ (p)$ in four or five rounds. Under the fish market treatment, the result was worse; the simulated players with the REL-1 model chose the $S\ (w) = D\ (p)$ point less than once. Simulated players with the REL-1-7 and REL-2 models chose this point at most once, while human subjects realized $S\ (w) = D\ (p)$ in three or four rounds. It is obvious that the D-REL-1 and D-REL-2 players choose the $S\ (w) = D\ (p)$ point more frequently than the REL-1 and REL-2 players. Under the stock market treatment, in each of the last five rounds, simulated players chose this point. Under the fish market treatment, in the last five rounds, they choose such a point more than thrice. The incorporation of such a deterministic rule into the REL improves its descriptive power.

1.5 Discussion and Conclusion

We arrive at the following results: 1. The REL model does not reproduce the behavior of human subjects very well. 2. The model that incorporates the $S(w) = D(p)$ rule to the REL model reproduces this behavior very well. 3. The model that limits the number of prices reproduces it well as compared to the model that does not. 4. The Best response strategy reproduces it well but lacks the reproduction of the *two-stage searching*.

Firstly, the REL neither exhibits the *two-stage searching* nor does it converge to a locally or globally optimal price set. This result may be affected by the fact that the number of rounds is short for simulated players. In addition, the result under the fish market treatment may not converge anywhere within twenty rounds, because the dead inventory loss is detrimental to learning. Secondly, our simulation results suggest that subjects use various rules such as the $S(w) = D(p)$ rule in addition to the REL. Thirdly, the simulation result suggests that the human subjects eliminate the absurd alternatives in advance. Finally, although the both MSD values of the Best response strategy are small, this strategy lacks the fine search, which is obvious because complete information about supply and demand functions is assumed unlike the case of our human experiment. Since human subjects understand this information with progression of the rounds, this strategy explains the behavior of subjects in the later rounds better than in the initial rounds.

We now conclude our paper. From the perspective of agent-modelling researches, laboratory experiments are very helpful in validating a model of economic actors. Agent-modelling would be one significant tool to magnify results obtained from laboratory experiments in terms of time scale and market size. Such estimated simulation would be helpful in deepening our understanding of human decision-making and designing market institutions.

References

1. Grosskopf, B., 2003 Reinforcement and Directional Learning in the Ultimatum Game with Responder Competition *Experimental Economics*, vol.6, pp.141-158
2. Erev, I. and Barron, G., 2003 On Adaptation, Maximization, and Reinforcement Learning Among Cognitive Strategies Technion Working Paper
3. Erev, I. and Roth, A., 2001 Simple Reinforcement Learning Model and Reciprocation in the Prisoner's Dilemma Game *BOUNDED RATIONALITY: The Adaptive Toolbox*, pp.215–231
4. Feltovich, N., 2000 Reinforcement-Based vs. Belief-Based Learning in Experimental Asymmetric-Information Games, *Econometrica*, vol.68, pp.605–641
5. Iwasaki, A., Imura S., Oda, S. H., Hatono, I., and Ueda, K., 2003 Does Reinforcement Learning Simulate Threshold Public Goods Games? A Comparison with Subject Experiments, *the Special Issue on Software Agent and its Applications of IEICE Transaction on Information and Systems*, E86-D, No.8, pp.1335–1343
6. Ogawa, K., Koyama, Y., and Oda, S.H., 2004(a) "A Middleman in An Ambiguous Situation: Experimental Evidence", *Journal of Socio-Economics*, in press.
7. Ogawa, K., Iyori, K. and Oda, S.H., 2004(b) Price Competition Between Middlemen: An Experimental Study *Gaming, Simulations, and Society: Research Scope and Perspective, Shiratori, Arai, and Kato eds.*, pp.59-68,
8. Roth, A. and Erev, I., 1995 Learning in Extensive-Form Games: Experimental Data and Simple Dynamic Models in the Intermediate Term *Games and Economic Behavior*, vol.8(1), pp. 164–212
9. Selten, R., 1998 Axiomatic Characterization of the Quadratic Scoring Rule, *Experimental Economics*, vol.1, pp.43–62
10. Spulber, D. F., 1999 *Market Microstructure -Intermediaries and the Theory of the Firm-*, Cambridge University Press.

Why should I Believe Your Signal? An Application of Adaptive Learning Model of Hypergame in Competitive Signaling Interaction

Pri Hermawan and Utomo Sarjono Putro

School of Business and Management, Bandung Institute of Technology, Jl. Ganesha No. 10, Bandung 40132, Indonesia

Abstract

This paper aims to get a better understanding of the dynamic process in the competitive signaling interaction, by studying the process from hypergame perspective. Competitive signaling interaction in this paper is about under what condition a company should believe its competitor's announcement in the interaction between a company who sends a price increase signal and who can choose between telling the truth and bluffing, and a company who receive the signal which has to decide whether or not to believe the sender. First, the concept of competitive signaling as a hypergame situation is identified. Then, the adaptive learning model of hypergame is described. Based on the model, this research proposed a new learning procedure to support a receiving firm, who faces the uncertainty of sender's motive, to learn the objective situation. Finally, by examining the simulation results, this research point out that a receiving firm which is supported with the learning procedure could learns the sender's motive, makes a better decision, and takes an action appropriate with it.

Keyword: marketing, price increase signal, hypergame, genetic algorithm, learning.

1 Signaling Interaction as Hypergame Situation

Previous research on marketing (Rymon et al. 1995) has investigated the strategic interaction between a sender and a receiver in a price increase signaling interaction where the sender has to choose whether or not to bluff about an intended price increase. In this research, the receiver has to choose whether or not to believe the sender's signal on the intended price increase. The objective of sending a truthful price increase signal is to encourage the competitor to cooperate by matching the new price given that direct face-to-face communication between competitors about pricing is prohibited. They used the conceptual framework presented in Truth Game (Brams 1985). In that game, a sender chooses between telling the truth and bluffing, and after the sender has made choice, a receiver decides whether to believe or not believe the signal. It is assumed that the receiver cannot tell for certain whether the sender was truthful or not.

However, that research still assumes that each player know the objective situation. In contrast, hypergame (Bennet 1986) argues that each player involved even in the same situation may see it differently and subjectively. Previous research (Putro et al. 2000) have built an Adaptive Learning Procedure that can be used in hypergame situation by using genetic algorithm (GA). Our research's focus is to help a receiver to learn about the sender motivation in hypergame situation. Concept of hypergame is adopted to describe the complex situation. It represents an agent's internal model of the situation with a subjective game. In hypergame, the objective situation is called the objective game, but agents may have misperceptions about it and learn it as subjective games. The model of learning for a group of agents that is investigated in this paper is illustrated in Figure 1.

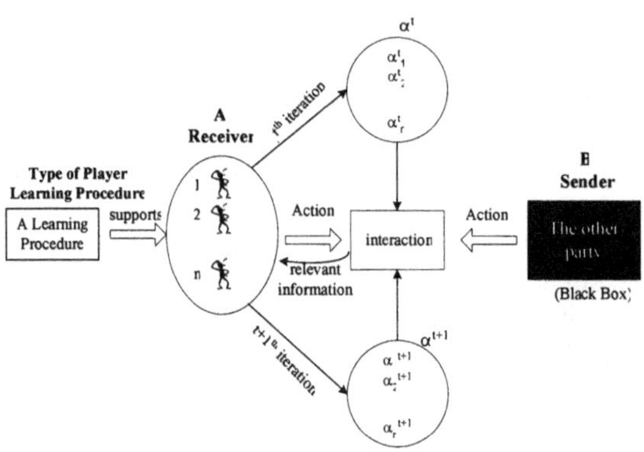

Fig. 1. Adaptive learning model of the hypergame situation

This research will capture the two possible player types: opportunistic and non-opportunistic. An opportunistic player seeks to take advantage of the other player, whereas a non-opportunistic player does not. Opportunism involves an element of deception. Allowing each player to be one of two types lead to four different objective situations or Scenario 1 to Scenario 4 respectively: (i) an opportunistic receiver interacting with an opportunistic sender, (ii) a non-opportunistic receiver interacting with an opportunistic sender, (iii) an opportunistic receiver interacting with a non-opportunistic sender, and (iv) a non-opportunistic receiver interacting with a non-opportunistic sender. This paper will use the adaptive learning model as a framework to analyze this competitive signaling interaction. This research proposed a new learning procedure, called Type of Player Learning Procedure (TPLP), to support receiver to be able to learn sender's motive, which is opportunistic or non-opportunistic.

2 Type of Player Learning Procedure

The activity of TPLP is described as follows. The first stage is initialization. This stage will generate randomly an initial population of n internal models, each of which is associated with a subjective game of agent $i \in A$. Each internal model is encoded by a bit string (or a chromosome). Second stage is action choice. In this stage, given the internal model and some information about the other group, each agent $i \in A$ plays a game with partner $j \in B$. Partner is selected by random. The way the agent chooses an action is defined by a learning procedure adopted by the group. Third stage is to revise the current population of internal models by GA.

2.1 Encoding

Elements of receiver strategy (S_A), namely "Believe" and "Not Believe", are represented with a_1 and a_2 respectively; while the elements of sender strategy (S_B) (i.e., "Tell Truth" and "Bluff") are represented with b_1 and b_2 respectively. With each $(a_x, b_y) \in S_A \times S_B$ is associated a function $g_i(i \in A\{1,2,...,n\})$ called genes defined by: $g_i : \{(a_x, b_y) | (a_x, b_y) \in S_A \times S_B\} \to \{1,2,3,4\}$, in which such that for each $a_x \in S_A$, $g_i(a_x, b_y) = \pi_{Bi}(a_x, b_y)$.

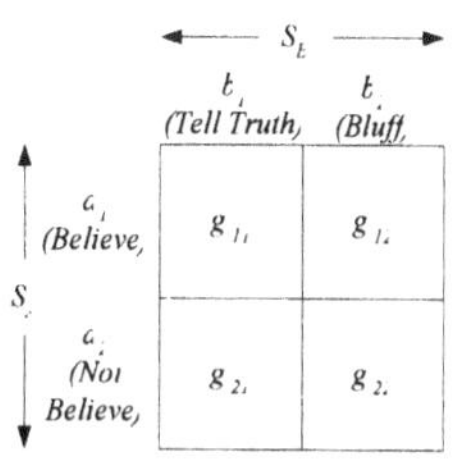

Fig. 2. Matrix form of chromosome of agent $i \in A$

2.2 Action Choice

An action choice represents how an agent chooses an action according to his/her internal model or some information about other group. Let N_i be a set of subjective Nash equilibria in internal model of agent $i \in A$. For each $i \in A$, i adopts the subjective Nash equilibrium action as follows:

- If $N_i = \emptyset$, agent i chooses an action from S_A randomly. In other words, if there is no subjective Nash equilibrium in his/her current internal model, agent i adopt random action, since his/her current internal model gives no information about the equilibrium.
- If $N_i = \{(a_m^*, b_m^*)\}$, agent i chooses a_m^* as his/her current action. That is, if there is only one subjective Nash equilibrium in his/her internal model, agent i applies the subjective Nash equilibrium action.
- If there are more that one element in N_i, agent i selects $(a_m^*, b_m^*) \in N_i$ such that for all $(a_x, b_y) \in N_i$, $\pi_A(a_m^*.b_m^*) \geq \pi_A(a_x, b_y)$, and then agent i chooses a_m^* as his/her current action. If there are more than one subjective Nash equilibria in his/her current internal model, agent i chooses the most preferred subjective Nash equilibrium action. π_A is agent i's preference over any outcome in $S_A \times S_{Bi}$ in ordinal utility function.

2.3 Fitness Evaluation

Fitness evaluation is to represents how an agent determines the fitness of his/her internal model by evaluating its consistency to some admissible information so far, and/or current payoff achieved. Let A adopts TPLP, then each agent i in the group evaluates fitness score of his/her internal model by a fitness function called f. Agent i evaluates internal model by taking into account his/her current payoff, and the frequency distribution of B's different responses to each action of S_A so far. The most frequent response to an action of S_A will be considered by i as the most preferred response of B to the action.

2.4 Selection

Selection is to choose an internal model, called parent, from the population of n internal models, when each agent begins to revise his/her current internal model. The learning procedure use a selection method called "roulette wheel" based on the ratio of fitness score of each internal model to the total fitness score of all the present internal model, that is described as follows:

- Generate a random number between 0 and 1, called *Rand.*
- Let $Fit(i)$ be fitness score of i's internal model, then calculate $RatioFit(i) = \frac{Fit(i)}{\sum_{k=1}^{n} Fit(k)}$, for each agent $i \in A = \{1,2,...,n\}$.
- Select internal model of $j \in A$ as a parent if $\sum_{k=1}^{j} RatioFit(k) \geq Rand$ and there is no $x \in A$, $x < j$ such that $\sum_{k=1}^{x} RatioFit(k) \geq Rand$.

After doing selection, each agent in the group begins to create children by reproduction or crossover. The group adopts a mechanism that decides whether a pair of perceptions adopts crossover or reproduction as follows. First, it generates a random number $u \in [0,1]$. Then, if $u < p_c$, the pair adopt crossover; otherwise they adopt reproduction.

2.5 Reproduction and Crossover

Reproduction is used by decision makers to produce new internal models simply by copying the parents (or internal models chosen by them in the selection). This research uses single point crossover. The procedure is as follows:

- Select cut point, namely the gene g_{2c}^{r} as the cut point of C_r and C_s respectively.
- Combine the parents; C_r and C_s are combined by exchanging each other genes located in the right hand side of the cut point.
- Revise the combination results n turn by the steps below:
  - Adjust perception about S_B implied by the combination results with the set of different actions that was applied by *B* so far.
  - Let an agent $i \in A$ interprets the current interaction result with his/her partner as a cell (l^{th} line, c^{th} column), then in this step, *i* changes the value of the gene in the cell into 4. It means that agent *i* assumes that the other group has the highest preference on the current result.
  - Rank the values of the genes in each chromosome, and change the value of each gene in the chromosome into its rank

3 Simulation

The following is the simulation procedure of the learning process in the competitive signaling interaction.

- Step 1 Initialization: Execute Stage 1 of the learning model for both the groups simultaneously.

- Step 2 Playing games: Execute Stage 2 of the learning model for both the groups simultaneously.
- Step 3 Termination: If the population of internal models of each group has converged, or the maximum number of iteration has been reached, then STOP. Otherwise, go to Step 4.
- Step 4 Update: Execute the Stage 3 of the learning model for receiver.
- Step 5 Games repetition: Go to Step 2 for receiver and to Step 1 for sender.

Each iteration of the simulation procedure is called a generation. The entire set of generations from the beginning until termination is called a run. Each run produces a final internal model for each group. A final internal model is an internal model on which a population of internal models of a group has converged. Based on the results of pre-simulation, the parameters for the simulation of the learning process are set as follow. The number of agents in each groups (n) is 30, the maximum number of iterations is 500, the probability of crossover (or p_c) is 60%, and w_1 and w_2 are 7 and 21 respectively.

4 Simulation Results and Discussions

The simulation results for each scenario are described in Figure 3 to Figure 6. In this research is assumed that receiver's perceptions do not know the sender's type. So, the decision maker has to learn it and take an action appropriate with the learning results.

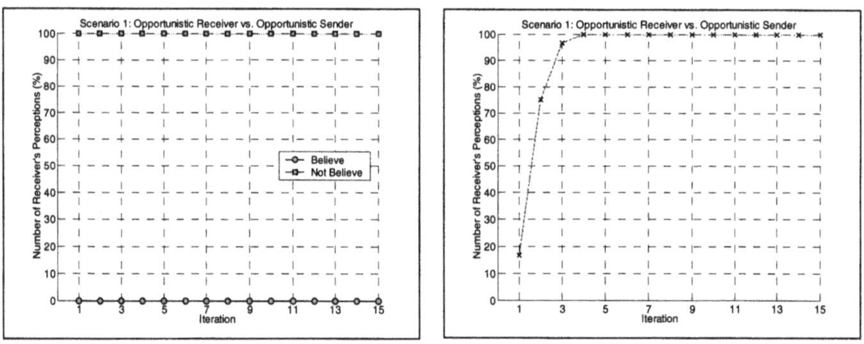

Fig. 3. Graph of number of receiver's perceptions which choose an action in Scenario 1 (left). Graph of number of receiver's perceptions which successfully learn that sender is opportunist (right)

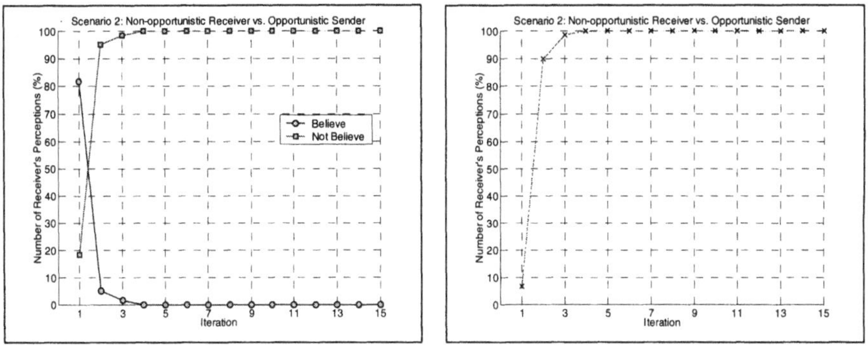

Fig. 4. Graph of number of receiver's perceptions which choose an action in Scenario 2 (left). Graph of number of receiver's perceptions which successfully learn that sender is opportunist (right)

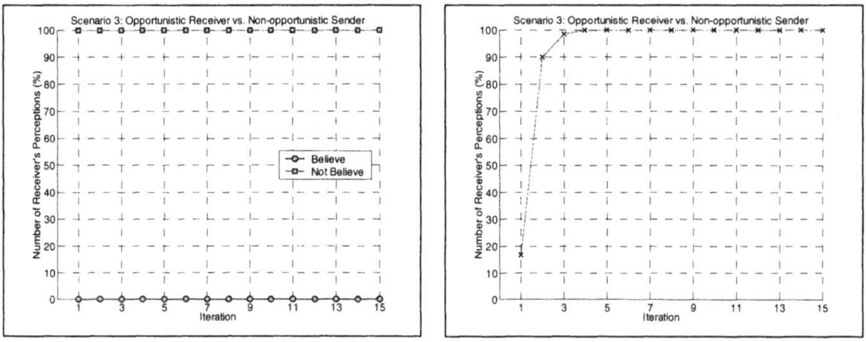

Fig. 5. Graph of number of receiver's perceptions which choose an action in Scenario 3 (left). Graph of number of receiver's perceptions which successfully learn that sender is non-opportunist (right)

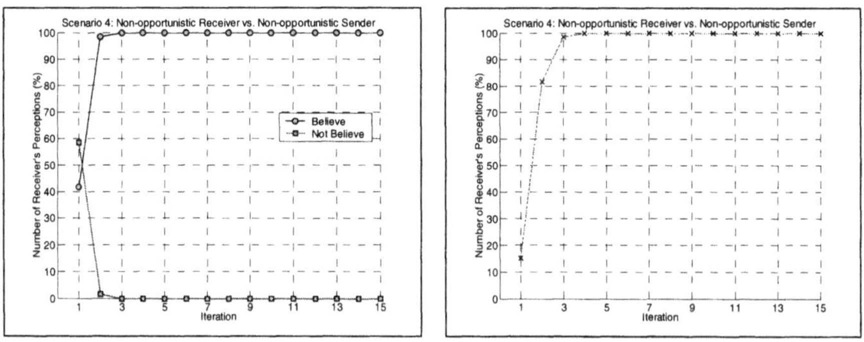

Fig. 6. Graph of number of receiver's perceptions which choose an action in Scenario 4 (left). Graph of number of receiver's perceptions which successfully learn that sender is non-opportunist (right)

The results show that in all scenarios, receiver, which is supported with TPLP, can learn the sender motivation, opportunist or non-opportunist. And based the learning, receiver will take an action appropriate with it. In Scenario1, receiver will always not believe the sender's signal, due to the learning result and its own opportunistic player's type. In Scenario 2, the learning procedure can guide receiver to escape from being taken advantage by the opportunistic sender. The receiver's action will become not believe the signal. In Scenario 3, although receiver can learn that sender is non-opportunistic, but because its own opportunistic player's type, the receiver will always not believe the sender's signal. In Scenario 4, the receiver can learn that sender is non-opportunistic and sending a truthful signal. Appropriate with it, receiver believes the signal, and a cooperative situation is achieved.

5 Conclusions

In general, this research shows that receiver which supported with TPLP able to learn sender's type and choose an appropriate action. Receiver's behavior to evaluates their perception on current situation based on best response to preceding partner's action can be used to guide a non opportunistic receiver to escape from being taken advantage by an opportunist sender When both receiver and sender has non-opportunistic motivation, this behavior guides the interaction to be stable in the cooperative relationship.

References

Bennet PG, Cropper S, Huxham C (1989) Modeling Interactive Decisions: The Hypergame Focus. In: Rosenhead J (ed) Rational Analysis for A Problematic World. Wiley, Chichester

Brams SJ (1985) Superpower Games: Applying Game Theory to Superpower Conflict. Yale University Press, New Haven and London

Putro US, Kijima K, Takahashi S (2000) Adaptive Learning of Hypergame Situation by Using Genetic Algorithm. IEEE Transactions on Systems, Man, and Cybernetics – Part A: Systems and Humans 30: 562–572

Rymon T, Eliashberg J, Robertson TS (1995) When Should You Believe Your Competitor's Announcement. (Institute for the Study of Business Market Report 8-1995, The Pennsylvania State University)

Cooperation is not always so simple to learn.

M. Mailliard, F. Amblard, C. Sibertin-Blanc, P.Roggero

IRIT, LEREPS-CIRESS – Université de Toulouse 1
21, allées de Brienne, 31 042 Toulouse Cedex – France
mmaillia@univ-tlse1.fr, sibertin@univ-tlse1.fr, famblard@univ-tlse1.fr, progger@univ-tlse1.fr

Abstract. In this paper, we propose to study the influence of different learning mechanisms of social behaviours on a given multi-agent model (Sibertin-Blanc et al. 2005). The studied model has been elaborated from a formalization of the organized action theory (Crozier and Friedberg 1977) and is based on the modelling of control and dependency relationships between resources and actors. The proposed learning mechanisms cover different possible implementations of the classifier systems on this model. In order to compare our results with existing ones in a classical framework, we restrain here the study to cases corresponding to the prisoner's dilemma framework. The obtained results exhibit variability about convergence times as well as emergent social behaviours depending on the implementation choices for the learning classifier systems (LCS) and on the LCS parameters. We conclude by analysing the sources of this variability and giving perspectives about the use of such a model in broader cases.

1. Introduction

The way social actions are coordinated by and among social actors has been a source of inspiration for many theories, as game theory, in very different domains such as economics, ecology (Dugatkin 1984), sociology or even in psychology. In a larger perspective, we choose as a research project to investigate the sociological theory of organized action proposed by (Crozier and Friedberg 1977), on the one hand to improve this discursive theory by proposing its formalization and on the other hand to apply it to model different socio-organizationnal phenomena. The work conducted on this project resulted in a proposed formalization of this theory (Sibertin-Blanc et al. 2005), a meta-model, that we briefly expose in the section 2.

Having this meta-model as a first milestone, we are then searching to improve it including social learning mechanisms enabling actors to act rationally. So we focus on LCS as a simple learning mechanism able to represent social learning. Different implementation choices being realistic, we decided to study those alternatives in order to make up our mind. We present those alternatives in section 3 as well as their sociological interpretations in the frame of the organized action theory.

Given those learning mechanisms, we are searching to understand in each case, which collective strategies could emerge and why. We then proceed to experimentations on each alternative in section 4. In order to make things understandable, we choose to make vary parameters corresponding to the sharing of resources in a perticular organization from which we generalize the famous prisoner's dilemma game.

The experimentations exhibit a variability in the emergent collective behaviours depending on the chosen parameters as well as a phase transition at the tipping-point corresponding to the individual transition from dependency on the needed resources to control on those resources. Results are given section 5. In section 6, we provide conclusions concerning the comparison of the different learning mechanisms. We conclude by giving the steps envisaged to follow up this work.

2. Formalization of the Organized Action sociological theory

A formalization of the Sociology of the Organized Action (SOA) leads to consider that constitutive elements of a social system are the *Actors*, the *Relations* and the *Resources*. In this paper, we will only focus on the Actors and the Resources (Fig. 1.).

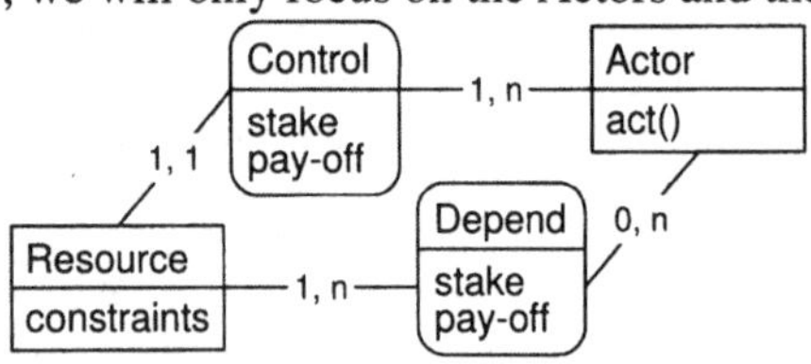

Fig. 1. Entity/Association Model of the structure of a social system in the frame of the SOA.

Actors could be defined as the entities of a social system. A Resource represents an object or a mean, needed by one or many Actors to achieve their action within the organization. Each Resource is linked to an Actor who controls it, and other associated ones being dependant on it. Each actor puts stakes on those Resources and receives in return a pay-off for each one of the resources he is linked to. The actor who controls a Resource decides of the distribution of the pay-offs among the Actors who depend on it and therefore influence their action capability. Every Actor controls one or more Resources and then possesses some freedom to act.

The pay-off corresponds to the quality of the Resource availability; more or better the usability of the Resource by an Actor, higher is its pay-off for this Resource.

The distribution of pay-offs and stakes on numerical scales enables, applying simple operations, to aggregate those values in synthetic and significant values. One can graduate the stakes on a scale between 0 and +10, and the pay-offs with the correspondence –10 to +10. As evidence, these numerical values just enable to perform comparison among them. To do so, we have to normalize the sum of the actors' stakes and then attribute the same amount of stakes to each actor for him to distribute on the relations he participates to. This normalization comes down to grant the same investment to each actor, i.e. the same possibility of personal implication in the social relations game.

A particularly significant value is, for each actor, the sum on the whole set of Resources he is involved in, of a combination between his stake and the resulting payoff he receive. We name this value the actor's *satisfaction* (rather than utility because it is more linked to a context of bounded-rationality). It expresses the possibility for an actor to access the resources he needs in order to achieve his objectives, and then the means available to him considering these objectives. A linear version consists in considering the sum, on every relation he is involved in, of the stake by the pay-off:

$$Satisfaction(a) = \sum_{r/\ a\ participates\ to\ r} stake(a, r) * pay\text{-}off(a, r) \qquad \textbf{(1)}$$

To attain or preserve a high level for this satisfaction is a meta-objective for every actor, as this level determines his possibility to achieve his concrete objectives. The strategic characteristic of an actor's behaviour leads him, by definition, to aim and achieve his objectives and then to obtain an acceptable value (if not the optimum) for his satisfaction, that becomes the criterion for learning mechanisms we expose in the following section.

3. The learning mechanisms implemented

After the claim by (Conte and Paolucci 2001) among others for integrating so-called « intelligent » social processes in the agents when doing agent-based social simulations, several papers (Conte and Paolucci 2001, Flache and Macy 2002, Takadama et al. 2003) proposed some learning algorithms to be used by agents.

In this paper, we explore two models for social learning using the learning classifier systems, LCS (Holland et al. 2000), for the action selection. LCS are based on the learning of behavioural rules using a test-errors approach, reinforcing the rules depending on the results they produce in a given context. Recent works about reinforcement learning models exhibit that a reduced set of parameters and hypothesis may cover important hidden theoretical assumptions (Macy and Flache 2002). In order to advance prudently we decide to make their use explicit and to discuss the way they can be implemented.

Each model is a LCS without genetic algorithms nor bucket brigad retribution process. One can refer to (Sibertin et al. 2005) to access to the detail of the used algorithm. The main processes involved, namely retribution, oblivion and matching of the learned rules are respectively governed by three parameters:

- *reward* is a positive or negative reinforcement of the rules depending on Δ*satisfaction*;
- *oblivion* is a factor of reward mainly used to weaken the strength of each rules;
- *dmin* enables an agent to match a perceived situation with yet learned situation-action rules.

Finally, the election process performs the selection of the matching rule with the highest strength. If there is none, the covering process generates a random one.

As exprimed by (Molm 1991), many social scientists denote that satisfaction may be specific or global. We thus propose to study a specific and a global satisfaction model: *Specific Satisfaction CS* (SSCS) and *Global Satisfaction CS* (GSCS).

Our first definition of satisfaction as exprimed in equation (1) is in the scope of Molm's definition of satisfaction: "*cognitive evaluations in which actors compare actual to expected outcomes*". We use this for the GSCS model. For the SSCS each actor *a* associates to each resource *r* a *satisfaction* expressed as:

$$\text{Satisfaction}_{a,r} = \text{stake}_{a,r} * \text{pay-off}_{a,r} \quad \textbf{(2)}$$

The main difference between GSCS and SSCS results in a local or global feedback of actors' actions.

4. Experimentations conducted

In order to validate our model we propose to use a cross-validation, as proposed by (Takadama 2003), in a prisonner's dilemma (PD) game. Althougth Takadama's work has been a rich influence leading our work, we use another experimental protocol.

A general difference is to consider the simulator level analysis versus the social system level one. Firstly, this lead us to an exhaustive exploration of *oblivion/reward* and d_{min}, parameters on one side, and *stakes* parameters on the other one. This exploration goes beyond the constraints of the prisoner's dilemma and enables to situate the results in a wider area. Secondly, at the simulator level, we base the validation on a homogeneous behaviour whatever are the social system parameters, and at the social system level: we are searching for parameter values under which mutual cooperation in a PD game is the most frequent social behaviour, as in everyday life relationships. The last difference with Takadama work is that agents' representations are the same for both compared models

4.1. The Prisoner's Dilemma

The prisoner's dilemma was proposed by two mathematicians Merrill Flood and Melvin Dresher in 1950. It is exposed as a game where two players, the prisoners, have the choice to cooperate, *c*, or to defect, *d*. Players earn pay-offs depending on the choices of the both players, as shown in Table 1. If the two players cooperate they will receive the reward for the cooperation (R); if both defect they will be punished for the defection (P); and if one cooperates whilst the other defects, he is the sucker (S) and the other will earn the retribution of his temptation (T).

The dilemma is constrained by the fact that temptation is more profitable than mutual cooperation (*cc*), that pays more than punishment, that is more valuable than to be the sucker: T > R > P > S; and that the best collective strategy is *cc*, 2 R > T + S.

The classical PD game is of minor interest compared to its iterated version where each player can potentially apply different actions over time and where the pay-offs are summed up. The iterated version of the PD has been widely explored and exposed (Hoffman 2000, Delahaye 1992, Macy and Flache 2002) since 1984 Axelrod's tournament.

The expressivity of the SOA formalization does not directly match the PD game. So we will present here how we make a projection from our model to enter in a PD game context. This projection leads us to a generalized PD game.

Lets define a *2-actors, 2-resources Organization* within SOA formalization. Let be two actors, *1* and *2*, and two resources, r_1 and r_2, such that each actor i controls the resource r_i. We normalize the stakes for each actor so that their sum is 10. Let be $s_{i,\,rj}$ and $p_{i,\,rj} \in [0;10]$ respectively the stake and the pay-offs of an actor i for a resource r_j. Let be $give_i$ and $take_i$ the possible actions each actor i can exert on its controlled resource r_i. Lets now define the effect of an action *action* applied by the controler c of a relation r as $effect_r(action) = \{\Delta p_{c,\,r}, \Delta p_{d,\,r}\}$ such that $\Delta p_{c,\,r}$ and $\Delta p_{d,\,r}$ are respectively the pay-off increments of the controler actor c and the dependant actor d of the relation r. Let be $effect(give) = effect^{-1}(take) = \{-1,1\}$, so that we are in a zero-sum game.

	$give_1$	$take_1$
$give_2$	$sat_1 = s_{1,r2} - s_{1,r1}$ $sat_2 = s_{2,r1} - s_{2,r2}$	$sat_1 = s_{1,r1} + s_{1,r2}$ $sat_2 = -s_{2,r1} - s_{2,r2}$
$take_2$	$sat_1 = -s_{1,r1} - s_{1,r2}$ $sat_2 = s_{2,r2} + s_{2,r1}$	$sat_1 = s_{1,r1} - s_{1,r2}$ $sat_2 = s_{2,r2} - s_{2,r1}$

→

	c_1	d_1
c_2	R_1 R_2	T_1 S_2
d_2	S_1 T_2	P_1 P_2

→

	c	d
c	R R	T S
d	S T	P P

2-actors, 2-resources Organizations ***⊃ Sociology of Organized Action Model*** ⊂ ***Generalized PD games*** ⊂ ***PD game***

Table 1. SOA Model enables to define 2-actors, 2-resources Organizations which include all Generalized PD games which include the famous PD game.

Lets now define the *Generalized PD Games*. In a *2-actors, 2-resources Organization,* lets consider that satisfaction sat_i of each actor i is determined by the played couple of action $\{action_1, action_2\}$, such that there are four potential different satisfaction values for each actor. Lets assume a syntactic equivalence between $\{sat_1(give_1, give_2), sat_2(give_1, give_2), \ldots\}$and $\{R_1, R_2, \ldots\}$ and another one between $\{give_i, take_i\}$ and $\{c_i, d_i\}$ as shown in the two first tables of Table 1.. Let be the constraints: $R_1 > S_1 > T_1 > P_1$, $R_2 > S_2 > T_2 > P_2$, $R_1+R_2 > T_1+S_2$ and $R_1+R_2 > T_2+S_1$.

Fig. 2. *2-actors, 2-resources Organizations Matrix.* The X-axis and Y- axis respectively represents the relative autonomy[1] of *1* ($s_{1,\,r1}$) and *2* ($s_{2,\,r2}$). Generalized PD Games are ploted in grey or white, while PD Games are in white.

Lets define the classical *PD Game* as a *Generalized PD Game* respecting the following constraints: $R_1 = R_2$, $S_1 = S_2$, $T_1 = T_2$ and $P_1 = P_2$.

4.2. Experimental design

The simulations were conducted with the same complete experimental design for the both models (SSCS and GSCS).

A first set of parameters concerns the sociological model, as it is the *stake* of each actor for the relation he controls. Because of the normalization, we do not have to

[1] We can directly deduced the actors' relative dependency ($s_{1,\,r1}$ and $s_{2,\,r1}$) from their relative autonomy because of the stake normalization.

make vary the *stake* of the actor for the other relation. Moreover, the game being symmetric we only explore one half of the parameter space.

A second set of parameters concerns the LCS. We explore the d_{min} and the *oblivion/reward* ratio as follow. The d_{min} is a distance which is used to compare situation part of a rule to the perceived situation. The greater d_{min} the less thick is the exploration. The possible values we have chosen to explore for d_{min} are graduated on a logarithmic scale from 0 (maximum of one applicable rule) , 2^0,... to 2^5 (all rules are applicable). The *oblivion/reward* ratio is also essential because it permits to renew the rules whose situation part is matching the current situation. A high ratio value will conduct to a quick renewal of the rules, whilst at the opposite a low ratio will slow down the adaptation of the agent. The ratio value belongs to [0;1] and is incremented by step of 0.2. The reward is fixed at 5.

We have produced 50 runs within a maximum of 5000 steps for each parameter quadruplet {stake_of_r1_controler, stake_of_r2_controler, dmin, oblivion/reward}. For each model and for each parameter quadruplet we have observed many values (mean and standard deviation for the convergence time…) needed for the validation.

5. Results

How to read the five dimensions representation ?

In the following of this document we use a combinaison of two bidimensional matrices to present some observations in a single bidimensional matrix. The construction principle is to affect a matrix to an element of the other one. This enable us to represent a value from the social system point of view in function of the LCS parameters or the opposite. If we analyse a value under the social system point of view, the main matrix will be the *stake matrix*, else it will be the *LCS matrix* (X-axis represents *dmin*, Y-axis represents oblivion/reward). 3-dimentional matrices are the result of applying the mean of each sub-matrix of the five dimensions representation.

5.1. Results for the Specific Satisfaction Classifier System

The frequency of occurence of each action pair (*cc*, *cd* or *dd*) at convergence point is 1 for each Generalized PD Game whatever are the simulator parameter values. As an evidence of violation of our second validation criterium, we do not need more investigation with this model.

5.2. Results for the Global Satisfaction Classifier System

The GSCS seems to give a largest variety of results than the SSCS. As we can observe on Fig. 3.a the larger d_{min} the lower the convergence time. The *oblivion/reward* ratio also speeds up the convergence as it increases. We observe a phase transition for a value of *dmin* between 2 and 4. Another observation presents in the mean convergence matrix (Fig. 3.b.) is a distinct transition which appears in the generalized PD game area. Homogeneity validation[2] is true for dmin > 8 (Fig. 3.c).

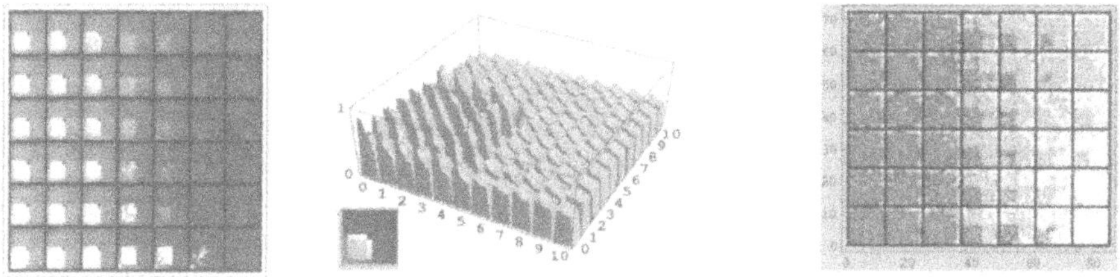

a. Mean convergence *LCS view* b. Mean convergence *Social system view* c. Behavior homogeneity (dark gray: invalidated)

Fig. 3. Convergence and LCS homogeneity. A transition appears in the generalized PD area (**b.**)

The results of the analysis of the frequency of occurence of each action pair at convergence give us that the *cc* pair emerges as more frequently as the *dmin* value is higher than 16 or lower than 2, and as the oblivion/reward rate increases; the *dd* pair emerges more frequently as *dmin* increases and as the oblivion/reward rate decreases; the *cd* pair emerges more frequently as the *dmin* value is upper 1 and lower 16.

We now can verify our second criterium, that is predominance of mutual cooperation. The criterium is true if *cc* are 20% more frequent than each other pair, that is when dmin is lower than 2 or equal to 32 and oblivion/reward is upper 0.4.

It is also remarkable that for dmin equals 32 and oblivion/reward ratio is upper 0.4 our first criterium about the homogeneity of the LCS behaviour is also validated, and thus the GSCS is validated for both criteria.

6. Discussion

We have proposed to compare the behaviour of two learning classifier systems in a generalized PD game in regard to a social system criterium, namely the predominance of the mutual cooperation, and to a simulation criterium, namely the LCS behavioural homogeneity at convergence point independently of the social system state.

As a matter of fact the SSCS model has no feedback on the whole system and thus is not suitable for agent adaptation in a mutual dependancy context. At the opposite, the GSCS enable values validating our criteria (oblivion/reward ≥ 0.6 and dmin=32).

As a first feedback of this study, and in regard to the transitions that we have exhibit, it would be interesting to relaxe our homogeneity criterium and to take into account that mutual cooperation could have a co-adaptation cost. The second feedback is that the frequency of occurrence for the *cd* action pairs is very high and seems to be due to an edge effect of the model that we have to study.

The explanation of the phase transition requires an interpretation of the behaviour of a *2-actors, 2-resources Organization* which will cover the classical interpretation of the PD Game. Such a work is out of the scope of this article.

---

[2] Standard deviation < 0.2 * mean (white on Fig. 4.c), or mean convergence < 50 (light grey).

7.Conclusion

As an evidence, the presented case is constraining considering the learning mechanisms that could be proposed. Following works by increasing the number of actors or by using different organisations of the relationships between actors and resources could lead to suggest other learning mechanisms. But in this case, as a result of our study, the GSCS is more adequate to our modelling purposes.

We have also to mention that the proposed model, as a formalization of the sociological theory of organized action, has a far broader spectrum of application than the one presented in the paper. Among others, we applied or derived this meta-model on the study of the emergence of territorial coalitions (Mailliard et al. 2005), to classical cases from the strategic analysis literature as the Trouville case (Mailliard et al. 2003).

Moreover, this interdisciplinary work even presented as the use of computer sciences as tools for sociological theories, benefits also to computer sciences as a source of inspiration in order to propose original coordination mechanisms among computational agents (Sibertin-Blanc et al. 2005).

References

Axelrod, R.: The evolution of cooperation. Basic Books, New York, (1984).

Conte, R., Paolucci, M.: Intelligent Social Learning. JASSS vol. 4, no. 1, (2001).

Crozier, M., Friedberg, E.: L'acteur et le système: les contraintes de l'action collective. Seuil (1977).

Delahaye, J. P., L'altruisme récompensé ? *Pour La Science*, 181:150-156, (1992).

Dugatkin, L.A.: Cooperation among Animals: An Evolutionary Perspective. Oxford University Press, (1997).

Flache A., Macy, M.W.: Stochastic colusion and the power law learning. Journal of Conflict Resolution, (2002).

Hoffmann, R.: Twenty Years on: The Evolution of Cooperation Revisited, Journal of Artificial Societies and Social Simulation (JASSS) vol. 3, no. 2, (2000).

Holland, J, All.: What Is a Learning Classifier System? LCS'99, LNAI 1813, 3-32 (2000).

Macy, M.W., Flache. A.. Learning Dynamics in Social Dilemmas. Proceedings of the National Academy of Sciences U.S.A. May 14;99(10):7229-36,(2002).

Mailliard,M.,Amblard, F., Sibertin-Blanc C.: Modélisation multi-agents pour la formalisation de théories sociologiques: Le cas de la sociologie de l'action organisée appliquée à l'étude de la dynamique du pays Quercy-Rouergue. In Proceedings of the SMAGET, France (2005).

Mailliard, M., Audras, S., Casula, M. : Multi Agents Systems based on Classifiers for the Simulation of Concrete Action Systems. In Proceedings of the 1st EUropean Workshop on Multi-Agent Systems (EUMAS), Oxford University, (2003).

Molm, L.: Affect and social exchange: satisfaction in power-dependence relations. American Sociological Review, vol. 56, (1991).

Sibertin-Blanc, C., Amblard, F., Mailliard, M.: A coordination framework based on the Sociology of the Organized Action, In Proceedings of the From Organization to Organization Oriented Programming in MAS, AAMAS, Utrecht University, (2005).

Simon, H.: The sciences of the artificial, MIT Press, 3rd edition (1996).

Takadama, K., and al.: Cross-Element Validation in Multiagent-based Simulation: Switching Learning Mechanisms in Agents. JASSS vol. 6, no. 4, (2003).

Methodology and Tools

Proposal of an Agent-Based Simulator for Education of Social Research

Hiroyuki Onishi[1] and Hajime Kita[2]

[1] Graduate School of Informatics, Kyoto University
[2] Academic Center for Computing and Media Studies, Kyoto University
kita@media.kyoto-u.ac.jp

Summary Interests for social research, for example, survey using questionnaire, has risen as means to understand situation of societies. Social research is quite common in business and public services, and hence it is important to train persons who conduct the social research in order to make the research more effective. On account of effectiveness of education, training through real practices of social research is ideal. However, from viewpoints of cost and ethics, opportunity of such training is limited. In this study, the authors propose to use an agent-based software that enables to experience the whole process of design, execution, and analysis of a survey by agent-based simulation, and develop a prototype software so that beginners can learn the theory and skills of social research through simulation.
key words : agent-based approach, social research, education

1 Introduction

One of the important usages of simulation is training. In complex and mission critical engineering systems such as aircrafts and nuclear power plants, training of operators in simulators becomes indispensable. Using technology of virtual reality and robotics, simulation-based training of medical operation has also been investigated. Further, with consideration of progress in personal computing, enhancement of distance education, requirement of cost reduction in teaching and ethical/environmental issues, substitution of chemical and biological experiments with computer simulation has been also studied and introduced in higher-education. While it has not yet been popular, utilization of simulation in education of social sciences will be an effective tool because of complexity of the social systems. Gilbert has proposed to use computer simulation for teaching sampling for social research using computer simulation in 1978 [5]. The agent-based approach combined with gaming-simulation expands possibility of education using simulation in social sciences. Taniguchi et al. has used an artificial market system in teaching market economics[3], and Kobayashi and Terano have developed a system for simulation of business plan aimed at education in business schools[4].

In the present paper, the authors discuss agent-based approach to simulation of social research, and education of social survey using questionnaire through computer simulation. Progress of personal computer, simulation technology and artificial intelligence enables simulation of social survey using software agents having complex behavior as population.

2 Social Research and Its Education

In contemporary society, importance of social research is increasing for business, public services and other activities. Survey using questionnaire is the most common method of social research, and it is used quite often. However, survey must be conducted carefully considering its statistical nature, bias caused by behavior of respondents, and other factors, and therefore training of the persons who conduct the social research is very important.

As for teaching social research, effectiveness of conventional teaching style such as lecture is quite limited, and to conduct a real survey, while it may be an ideal process to learn, is difficult because of its cost and ethical consideration. Utilization of computer simulation can be a good tool for teaching of social survey. Gilbert has proposed to use computer simulation for teaching sampling using computer simulation[5]. Progress of personal computer, simulation technology and artificial intelligence enables simulation of more complicated process of social survey using the agent-based approach. This paper proposes a prototype agent-based simulation system for teaching of social survey. It can be used to visualize statistical characteristics of sampling method, effects of biases involved by behavior of respondents and design of survey considering such effects. Figure 1 illustrates the concept of agent-based simulation of social survey for education.

3 Agent-Based Model of Social Survey

3.1 Requirements for System and Design Concept

Considering usage of a simulator in teaching, the simulator has to include elements to be taught for successful research. That is, the factors to be designed in survey such as, sampling of respondents and its statistical characteristics, design of questionnaire, methods of survey such as distribution media of questionnaire, should be experiences through simulation. It should be noted that, as an advantage of using simulation, difference of the results obtained by several trials with the same survey design can be also visualized.

Considering technical feasibility, in our simulator, we deal with a single choice from multiple alternatives as a type of question/answer. It is most

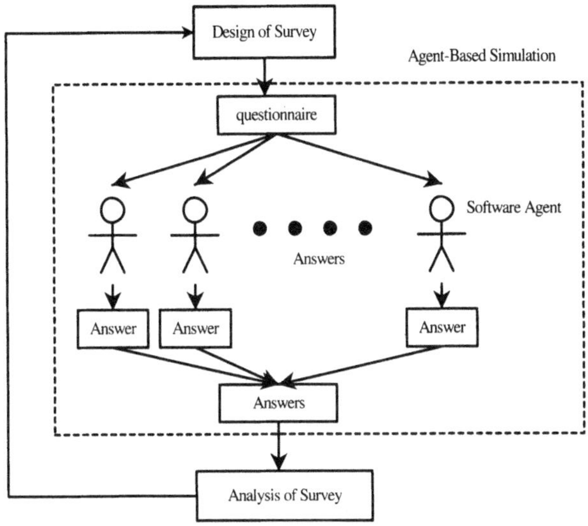

Fig. 1. Agent-based Simulation for Learning of Social Survey.

commonly used in survey. Each agent in the simulator has 'a ture answer' to each question in advance and returns it or another answer depending of the operation mode of the simulator, the behavioral model of the agent and design of the survey set by the user.

The following options in design of survey are considered in the simulator:

- Number of samples,
- Sampling methods, e.g., random, systematic, and stratified sampling,
- Methods of survey, e.g., face-to-face interview, survey via mail/telephone/web, and
- Usage of reward.

Design of questionnaire is also one of the central issues in teaching social survey. In our simulator, considering technical difficulty of dealing with free composition of question, we ask the learners to select questions from a set of questions prepared by a teacher. It enables to examine questions relating to various design issues to be taught into consideration.

3.2 Behavioral Model of Agent

In this simulator, respondents of the survey are modeled by software agents. Agents in the model work in two modes, i.e., 'the honest mode' in which the agents answer their true answers to the question, and 'the behavioral model mode' in which the agents change answers depending on the design and other factors in the simulated survey. We introduced the following two sorts of attributes as factors having influence on the agents behavior.

- Social Attributes: For example, gender, living place, age.
- Personal Attributes: For example, interest in the issue of survey, attitude to survey, patience to long questionnaire, sensitivity to reward.

Behavior of each agent is determined by combination of its attributes and design of survey.

In this study, we modeled behavior of the agent with two parameters, i.e., rate of cooperation p_a and rate of precise answer p_b. The rate of cooperation $p_a(i)$ of Agent i is determined by the factors of survey$\mathbf{f}_a = \{f_{a_1}, f_{a_2}, \cdots, f_{a_{k_a}}\}$ and agent's attribute $\mathbf{w}_i = \{w_{i,1}, w_{i,2}, \cdots, w_{i,n}\}$ as follows:

$$p_a(i) = \prod_{k=1}^{k_a} p_{a_k}(f_{a_k}, \mathbf{w}_i) \tag{1}$$

where $p_{a_k}(\in [0,1])$ is determined by f_{a_k} and $\mathbf{w}_i$.

As $p_{a_k}(i)$, we considered the following four factors:

1. Topic of survey: "Information Education in General Education Program"
2. Method of Survey: "Face to face interview", "Deliver and collect in class", "Deliver in class and return lately" and "Use Web"
3. Reward to respondent: "To use" or "Not to use"
4. Number of Questions

Concerning Factor 1, a value to $p_{a_1}(i)$ is set considering matching of the topic of survey and the agent's social attribute. For Factor 2, a table of values for $p_{a_2}(i)$ having entry of "attitude to survey" and "survey method" is prepared. Similarly, for Factor 3, a value table for $p_{a_3}(i)$ is prepared. Factor 4, which is a numerical factor, is treated by using sigmoid function whose parameters are x_0 and $a(i)$. Let the number of questions be x, and

$$p_{a_4} = \frac{1}{1 + \exp((x - x_0)/a)} \tag{2}$$

We use similar model for $p_b(i)$, and with $p_a(i), p_b(i)$, the agent acts as follows:

Step 1 (To decide to answer) : Agent i decides to answer the questionnaire with the probability of $p_a(i) \in [0,1]$ and proceeds to the next step. Otherwise it refuses to answer and quit.

Step 2 (To answer with fluctuation) : With the probability of $p_b(i)$, the agents chooses 'the true answer' and quit. Otherwise,

- If the question is in nominal scale, the agent chooses an alternative randomly. The answer includes 'DK (don't know)' and 'NA (no answer)' with the probability of p_{DKNA}.
- If the question is in the ordinal scale, the agent chooses alternative next to the true answer as follows.

- Let the alternative of the true answer be j, and alternatives next to the j be j^+(extreme side) and j^-(intermediate side). Agent i chooses j, j^+, j^- with the following probabilities:

$$p_j(i) = \tilde{p} + (1 - \tilde{p})(1 - p_c(i)) \tag{3}$$
$$p_{j^+}(i) = p_c(i)p^+ \tag{4}$$
$$p_{j^-}(i) = p_c(i)p^- \tag{5}$$

if j^+ doesn't exist,

$$p_j(i) = p_j(i) + p_{j^+} \tag{6}$$

where p^+, p^- are parameters and $\tilde{p} = 1 - (p^+ + p^-)$.

4 Implementation of Prototype System

4.1 Overview of the System

We have implemented a prototype system as a standalone application in Java using Swing. Figures 2 is the GUI of the prototype system. Design principle of this system is that the software provides the learners with only simulation of the survey. That is, no teaching material other than simulation such as explanation of the concepts and skills is included because it can be designed separately. The user interface of the prototype system consists of 7 parts:

Introduction Panel : It provides with the explanation of simulated survey.
Survey Options Panel : In this panel, the user sets options of the survey.
Questionnaire Design Panel : The user designs questionnaire by choosing a question set from candidate questions.
Sampling Panel : The user can execute sampling from population and gets a list of the sampled respondents.
Single Survey Panel : With this panel, the user can execute a single run of survey to the agents and obtain raw answer data from the respondents as well as simple statistics of the data.
Multiple Surveys Panel : With this panel, the user can try multiple runs of survey and examine statistical property of the results.
Graphical Presentation of Population : It shows the population graphically.

4.2 An Example of Teaching Plan

With the proposed simulator, the learners of social survey can learn fundamental statistical issues such as sampling and tests, and influence of behavior of the respondents actively. The following is an example teaching plan:

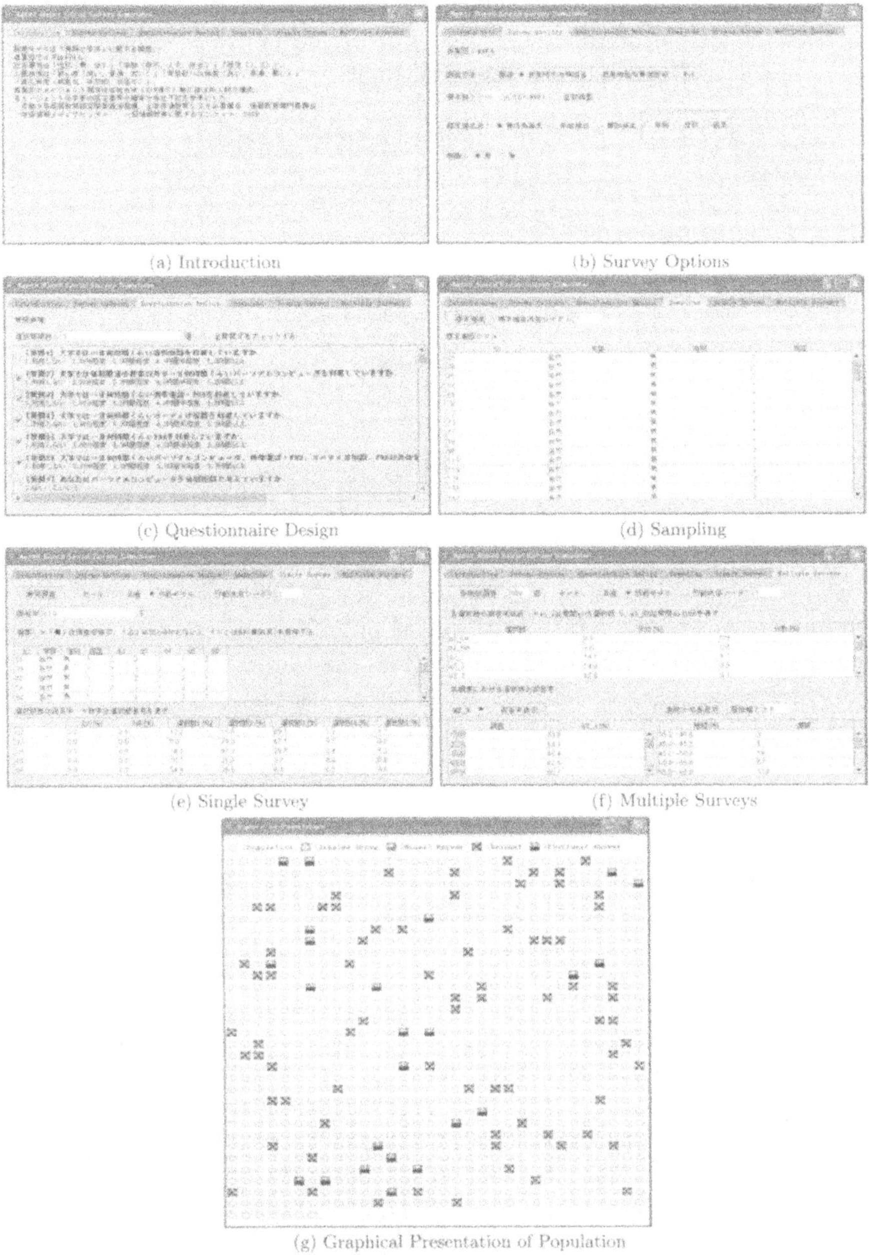

(a) Introduction (b) Survey Options

(c) Questionnaire Design (d) Sampling

(e) Single Survey (f) Multiple Surveys

(g) Graphical Presentation of Population

Fig. 2. User Interface of the Prototype Survey Simulator

- The instructor explain the related issues of social survey and the usage of the system with demonstration. In this example, the issue of distribution of sample ratio obtained by multiple sampling and comparison of the result with the population ratio. Comparison is also made for the honest mode and the behavioral mode.
- The learners, then try it by themselves.

Figures 3 and 4 are the results obtained by the simulator[3]. These figures shows the learners that the ratio obtained for sample is different from that of population while it is an approximation. It also changes by sample and distributes. It will be an good introduction to proceed topics of designing survey and analysis of the result such as statistical tests.

4.3 Evaluation of the System

We carried out peer evaluation of the prototype model and teaching plan. We have asked three university faculties having experience of social survey and/or teaching it to evaluate our prototype model and teaching plan. The comments obtained from the evaluators are as follows:

1. It is interesting because it provides learners with explorative data analysis.
2. It is interesting that we can see the honest answer and fluctuated answer from the same samples.
3. More realistic data should be used.
4. Improvement of the software in functions such as load/save survey setting, statistical analysis of the data, and graphical representation is needed.

The evaluators have said that they want to use it in the classroom if 1) Items 3 and 4 in the above comments are considered, 2) teaching plan is reexamined considering needed time, and 3) class is support by teaching assistants.

5 Conclusion

This paper proposes to use agent-based simulation in education of social research, and develops a prototype system of social survey simulator. Through peer evaluation of the system, it is expected to be an effective tool of teaching social survey. Improvement of the teaching plan reflecting the comments obtained by the peer evaluation, and refinement of the software for practical use in classroom are the subjects of the further study. From a technical point of view, refinement of the agent model and assistance in setting questions and agents' answers are also important subjects of this study.

[3] The simulator outputs the data in text, and the learners should handle it with the spreadsheet software. Design parameters of survey are as follows: Survey Method: Answer in Class, Samples: 100, Sampling Method: Random, Reward: None, Question: #51, Random Seed for Sampling:1, Random Seed for Agents:1

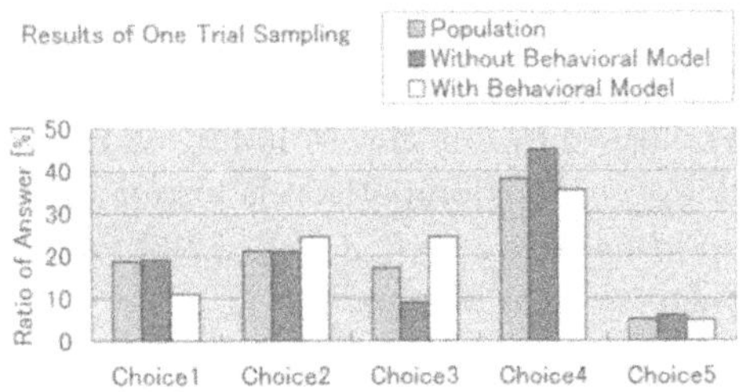

Fig. 3. Result of Simulation Run, Distribution of Selected Choices in One Sampled Survey Run.

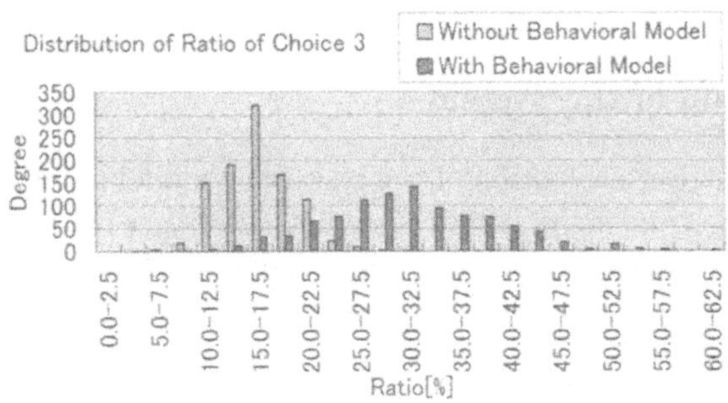

Fig. 4. Result of Simulation Run, Distribution of Selected Ratio of Choice 3 in 999 Survey Runs. (Mean of Population: 15.9% in honest mode, 31.2% in behavioral model mode)

Finally, the authors express their sincere gratitude for Prof. Hiroshi Deguchi and Dr.Yuhsuke Koyama of Tokyo Institute of Technology and Prof. Hachiro Iwai and Prof. Hiroyuki Matsui of Kyoto University for their valuable advice on this study.

References

1. Japan Certification Board for Social Researchers (`http://wwwsoc.nii.ac.jp/jcbsr/board2.htm`)
2. Robert Axelrod: The Complexity of Cooperation Agent-Based Models of Competition and Collaboration, Princeton University Press, 1997
3. K. Taniguchi et al.: A Report of U-Mart Experiments as an Educational Courseware, *Proc. AESCS'04*, pp. 193-198 (2004, Kyoto)
4. M. Kobayashi and T. Terano: Human-Agent Participation in a Business Simulator, in *Meeting the Challenge of Social Problem via Agent-Based Simulation*, Springer, pp. 91-106 (2003)
5. Gilbert, G.N., A simulation approach to teaching survey sampling, *Teaching Sociology*, pp. 287-293 (1978)

Political Science and Multi-Agent Simulation: Affinities, Examples and Possibilities

YAMAKAGE Susumu[1]
with Collaboration of
HOSHIRO Hiroyuki[2], MITSUTSUJI Katsuma[3], SAKAMOTO Takuto[3], SUZUKI Kazutoshi[3], and YAMAMOTO Kazuya[3]

1 Professor, the Graduate School of Arts and Sciences, the University of Tokyo
2 Research Associate, the Institute of Oriental Cultures, the University of Tokyo
3 Research Fellow, the Graduate School of Arts and Sciences, the University of Tokyo

Abstract. Domestic politics and international relations are fertile fields for the analysis using multi-agent simulation. Both theoretically and empirically, such a technique can serve the development of political science. To attract scholarly attention, some illustrative examples may be helpful. Herein, more or less realistic analyses of different types of interactions between different types of agents are outlined: (1) the White House during the Cuban Missile Crisis in 1962; (2) three-layered interrelationship between voters, politicians and political parties; (3) the long-lasted civil war in the Sudan; and (4) trade friction in interdependent world. Note that the same "simulator" was used in all those analyses. This multi-purpose "simulator," which has been developed in Japan, is introduced herein. Finally, some methodological possibilities are suggested.

1 Introduction

Agent-based simulation or multi-agent simulation (henceforth, MAS) is a promising method in analyzing political phenomena. Interactions between agents (either natural persons or nation-states) are essential in political behavior as compared with simple, price-taking micro economic behavior, for instance. In fact, numerous topics are suitable to the analysis applying MAS, and in recent years, an increasing number of researches have been conducted. On the other hand, however, they are produced by relatively a small number of researchers. In short, MAS is still a niche technique in political science. One of the obstacles that prevent a wider use of MAS is a doubt shared by many political scientists on the utility of simulation in general. However, the most serious obstacle seems the lack or short of skills to conduct MAS among political scientists. Compared with economics and sociology, political science is a discipline very distant from mathematical, numerical or quantitative methods, not to

---

The present paper was prepared for the Fourth International Workshop on Agent-based Approaches in Economic and Social Complex Systems/Annual Conference of Pacific-Asian Association for Agent-based Approach in Social Sciences, July 9-13, 2005, at Tokyo Institute of Technology. The author is grateful to Professor KANIE Norichika for giving me a chance to read the present paper.
Note that this Working Paper Series version is slightly different from the original one.

mention simulation. It is true that there are some interests in simulation, especially rapidly developing MAS, but it is yet to be disseminated. Few political scientists are familiar with programming technique to conduct MAS, while few collaborations with system/computer scientists has been attempted by political scientists.

We launched a research project in order to overcome such weak-points of most political scientists. Our approach is not to enhance computer literacy of political scientists, but to make computer technology closer and friendlier to them (more precisely to us including myself). Specifically, our aim is three-fold. First, to develop a user-friendly and multi-(if not omni-)purpose platform for MAS. Second, to analyze different types of social order through MAS. Third, to promote political analyses using MAS, and to encourage political scientists to try MAS in their respective research. Our endeavor is under way, and a few results can be shown.

2 Affinity of Political Phenomena for Multi-Agent Simulation

Politics is associated with power, control and authority. A ruler orders subordinates in a state; a leader directs followers in a company; major powers dominate smaller nations in international society. Such regulatory characteristics found in politics seem remotely related with the subject of MAS. There is another aspect in politics, however. It is the interplay between actors, each of which attempts to influence others; it is an art of collective choice; and it deals with social dynamics between chaos and order. There are a lot of cases that are suited for the analysis applying MAS. While Johnson(1999) conducted an extensive survey of simulation modeling in political science, the following attempt is to identify some promising fields of political science for MAS, and to point out relevant works with special reference to the outcome of our research project and to works done in Japan.

2.1 Political Theory

Emergence of social order is the central theme of political theory. Once the God-given order of *respublica Christiana* disappeared some centuries ago, alternative logics, more or less rational, have been formulated until even today. Namely, when the order was badly needed during the turbulent period of early modern Europe, a set of theories was developed by such philosophers as Hobbes, Locke and Rousseau, to name a few, which is now classified into the Social Contract Theory. Those classical theories are contending with one another, and Yamamoto(2002) evaluated the compatibility between their premises and dynamic outcomes based on them, by reinterpreting Hobbesean and Rousseauistic logics by MAS.

In general, political theory is to be developed by MAS-related studies. Within political science, political theory is most (in fact, purely) formal. Therefore, it is easy to translate the argument into the computer model. More importantly, political theory employs a bottom-up approach in the sense that it is concerned with the linkage between interactions among individuals, who are equal in nature, on the one hand, and the pattern of social order that emerges from the interaction on the other. Furthermore, MAS-based studies are free from oft-mentioned criticism against MAS on the short of empirical relevance because political theory is so axiomatic that only theoretical relevance matters. The pioneering analytical works on the tit-for-tat strategy in the prisoners' dilemma (PD) games by Axelrod(1984, 1997) can be classified into this category, although his methods and conclusions were later criticized, for instance, by Shimizu(1996) and Yasuda(2001).

2.2 Democracy

Democracy is an idea, and an institution as well, that formulate more or less binding social rules by transforming conflicting opinions or positions borne by different individuals into one authoritative conclusion on various issues. Political scientists have been concerned with various types of questions on how to make democracy work democratically. In this regard, one of the fundamental questions that are relevant to MAS is how to better represent a variety of interests borne by numerous citizens. The well-studied paradoxes of voting are classical and ideal examples demonstrating that the way of bottom-up aggregation of individual preferences affects social/collective outcomes. The search for better election systems has been another important subject that connects agents (voters) and collective will. Recent works include (Bhavnani 2003), (Bissey 2004) and (Kottonau and Pahl-Wostl 2004).

The public opinion plays important roles in democratic politics, and is certainly a promising subject to study by MAS. Based on the social impact theory (Latane and Wolf 1981), Nowak et al.(1990) simulated the change in individual attitudes and the formation of public opinion. In general, interactions between citizens and impacts of mass media are the two major subjects of concern.

2.3 Nationality and Ethnicity

American independence of England and French Revolution trumpeted the beginning of the age of nationalism. Since then, the politicization and mobilization of populations has become the main theme of politics. Nationalism involves thousands and millions of individuals in political arena. Culture and collective identity is brought in. In nation-states, ethnicity often plays a centrifugal role in domestic politics, and sometimes leads to ethnic conflicts. Civil wars are thus concerned with nationality and/or ethnicity of the people.

Politics of collective identity is the subject where MAS is widely applied. Axelrod(1997), Cederman(1997), Lustick et al.(2002) and Srbljinovic et al.(2003) show different but related approaches to analyze collective identity, and influenced following studies. Mitsutsuji(2002) illustrated the relationship between prevailing types of political movement and the geographical distribution of ethnic communities. Yamamoto(2004) extensively dealt with theories of nationalism, and evaluated the relevance of each theory based on MAS of cultural dynamics of populations.

2.4 International Relations

International Relations may be most suited for MAS-based studies. By definition, there is no single authoritative power in international society, and international affairs are the outcome where nation-states interact with one another so as to make alliance, war, and regimes. In fact, probably the earliest attempt of MAS in the field of political science was planned in the 1960s by Bremer and Mihalka(1977). The idea was further developed by Danno and Tanaka(1992) and Yamamoto et al.(2002). Those studies are concerned with the stability of balance of power, and the probability of the emergence of the empire. There are of course other types of MAS-based studies in International Relations. Yamamoto(2003) extensively surveyed and discussed the application of simulation techniques including MAS in the study of international relations.

Among various possible subjects, dynamics of international regime seems most promising. For, international regime is formed from the bottom rather than given by a supranational authority. The formation of regime, and the maintenance, is often related to the existence of the hegemonic power which can take lead and pay cost, but it is certainly the outcome of interactions between independent agents. In MAS-related studies, Kimura and Oda(2002) discussed the desirability of different types of regime in managing the greenhouse effect gases, and Hishiyama(2002) dealt with private regime of transnational telecommunication.

2.5 Some Illustrative Examples

As shown above, there are various fields of political science that are particularly appropriate for the MAS-related analysis. Needless to say, the above-mentioned fields are by no means exhaustive, and MAS can be employed in other fields, too. On the other hand, the application of MAS to political science will be promoted more easily if some illustrative examples are shown. They may provide prototypes of related studies, or may enhance imagination and insights toward MAS.

In this paper, four examples are outlined as shown below. Different types of MAS are intentionally selected from the models we have been constructing in the on-going research project so that a wide range of applicability of MAS can be shown. While the description is sketchy, they will hopefully suffice to illustrate the potentials of MAS.

3 Watershed at the Cuban Missile Crisis: Inside the ExCom

The Cuban Missile Crisis of 1962 was the most serious world affairs during the Cold War, and almost resulted in a nuclear war. The US Government chose "quarantine" –a kind of blockade of Cuba– out of six options, and succeeded in avoiding further escalation. Despite a huge number of memoirs and studies, it is unclear why and how that option was chosen. MAS may capture the characteristics of decision-making. Sakamoto and Hoshiro (2005) developed a model as follows.

3.1 Theoretical Background

To cope with the crisis, President John F. Kennedy convened the Executive Committee of the National Security Council (ExCom), whose members were the president's most trusted advisers (see **Fig. 3.1**). Various studies on this crisis decision have been accumulated, but few have successfully clarified the reason why "quarantine" was chosen. Even the most influential work by Graham Allison (1971, with Zellikow 1999), who attempted to explain the decision based on three alternative models, does not explain either frequent changes of individual ExCom members' views or the process of convergence to adopt quarantine.

In order to solve these problems, the "social comparison/ cultural value" theory was employed. It has been used explain the group polarization phenomenon in the field of social psychology (Brouw 1965; Stoner 1968; Myers and Lamm 1976; Olson and Suls 2000). The theory asserts that individuals in a given society, in addition to having their own views, recognize the majority's standard norms, and through group discussion and self-comparison, their views tend to shift toward a commonly accepted value. This logic was implanted into the model.

3.2 Overview of the Model

The main purpose is simulating the process of decision-making among ExCom members. Since there was one week for them to make decision, the model focuses on that period from the 16th of October when the CIA reported that the Soviet Union was placing offensive missiles in Cuba, up to the 22nd when JFK announced publicly that the US had decided to respond to the Soviet action with a naval quarantine of Cuba.

The model was designed to reflect the reality so as to allow the virtual ExCom members to change their views dynamically through their discussion. The individual agents' behavior was ruled by the above-mentioned "social comparison/ cultural value" theory. Assuming that there were six options from "Do Nothing" as the most cautious option to the riskiest, "Invasion", the simulation tested what opinions individual members originally had, and how those views shifted toward the majority's opinion throughout the one-week discussions.

3.3 Simulation Results and Implications

One of the most interesting results was that "Quarantine," which was chosen in reality, was as probable as a riskier option, "Air Strike" (**Fig. 3.2**). That is to say, the US Government could have adopted the policy that might lead to the war with more or less equal chance. Based on the realistic model, the counter-factual discussions can be done: for example, if the Secretary of State had been a dedicated pacifist, or if the Secretary of Defense had been a warmonger, how would the result have been different? Our experiments show that while the former would have led the group view to be more cautious, the latter would not have affected the result only delaying the convergence of their views.

A different model was further developed. The new model employs the cognitive map approach, and analyzes conversations between ExCom members more in depth. The analysis can be found in Sakamoto and Hoshiro (2006).

4 Electoral and Party Systems: Voters, Politicians and Parties

While the concept of liberal democracy is widely shared, democratic countries have different political institutions. In other words, there are different combinations of electoral systems and different party systems. There exist various arguments about the relationship between electoral and party systems. Mitsutsuji (2003, 2004) conducted MAS in creating an artificial political society as follows so as to explore this controversy.

4.1. Theoretical Background

Most famous and classical hypothesis is provided by Duberger(1954), which is the single-seated electoral district system generates two-party system. Empirical researches cast doubts about this causality. However, electoral system is said to be one of the most important factors that decide party system. It is assumed to influence political parties' and politicians' behavior strongly. What kinds of differences are brought about by the difference of electoral system?

4.2. Overview of the Model

In order to test the arguments, the Artificial Political Society (APS) was constructed in order to simulate dynamic interactions among voters, politicians and political parties. APS is made on the two-dimensional opinion space, which represents two issues of political preference: urban-based vs. rural-based along the vertical axis, and right wing vs. left wing along the horizontal axis. On the opinion space, three different types of agents (voters, politicians and political parties) interact with one another. Voters vote the politicians according to the similarity of preferences; elected politicians organize, join or transfer political parties; and political parties try to attract voters in order to win the election.

While they interact with one another, all move on the opinion space, i.e., change their political preferences, in either centrifugal or centripetal way depending on mutual relationship. Big swarms of voters may emerge around parties, which consequently consist of many politicians. In an opposite way, politicians' splits cause division and change in parties' opinions, which may dissolve a voters' swarm.

By dividing voters and candidates into electoral districts, electoral system is introduced into APS. For example, three types may be compared: the one district system of fifteen seats, the medium constituency system of five three-seat districts and the single-seat electoral district system of fifteen districts.

4.3. Simulation Results and Implications

The average number of parties under different conditions is shown in **Table 4.1**. It indicates that electoral system does not have significant impacts on the number of parties. Instead, other factors, such as easiness for politicians to change their party membership or easiness to form new parties, have more significant impacts on the number of parties.

The above findings do not mean that electoral system has insignificant impacts on politics. In APS, electoral system influences the distribution of politicians' opinions. **Figure 4.1** shows a typical distribution of politicians' opinions under the single seated district system and that under the one district system. Under the single seated system (left), because politicians have to fight for only one seat in each district, they tend to come closer to the center of voters' swarm. Under the one district system (right), on the other hand, because politicians who are not positioned at the center of voters may survive, politicians tend to spread over the swarm of voters, and their opinions vary to some extent.

5 Civil War and Mobilization: Real Sudan vs. Virtual Sudan

In failed/collapsed state, violent conflict is often associated with ethnic mobilization. Hence, the pattern of civil war tends to be determined by geographical setting of ethnic communities. The Sudan is a typical case: a multi-ethnic (racially, linguistically and religiously) country that suffered from civil war for more than two decades. Sakamoto (2004, 2005) examined the Sudan's uprisings through MAS by creating an artificial Sudan as follows.

5.1 Theoretical Background

Domestic conflicts are almost always accompanied by the mobilization process, a process in which inhabitants are politically mobilized by government and insurgent organization(s), thus dividing their country. Theoretical approaches to such conflicts have treated these divisions as given and static: they have seen conflicts within a context of some relationship between given social and/or cultural groups, each of which is derived from some common traits (e.g., ethnicity, religion, class). Many conflicts, however, are not so clear-cut, requiring dynamic analysis of the group formation process brought about by mobilization.

Axelrod's model of cultural dissemination (1997) and Cederman's model of nationality formation (1997) are pioneering attempts in the analysis of such phenomena. Reflecting the above theoretical concern, two kinds of simultaneous interactions during conflict, i.e. armed battles among organizations and mobilization of inhabitants by respective organizations, should be interrelated with each other.

5.2 Overview of the Model

A "virtual state", consisting of territory, inhabitants, and government, is to be constructed in computer, and rebellions take place somewhere in the territory stochastically. The inhabitants are spatially distributed over the territory and differentiated from each other according to several social/cultural traits. While fighting each other, the government and the insurgent organizations mobilize the inhabitants by manipulating "symbols" in order to acquire necessary supports and resources.

This general model is applied to the civil war in the Sudan since 1983. Based on ethno-linguistic and religious distribution maps, the virtual Sudan was constructed as an approximation of the real country (**Fig.5.1**).

5.3 Simulation Results and Implications

As for the Sudanese civil war, MAS should elucidate and connect the following two aspects: (1) the prolonged division of territorial rule between the northern-based government and the southern-based insurgent organization, the SPLA (Sudan People's Liberation Army), and (2) the political aspect of this division, namely the competition between the ethnically and religiously exclusive government and the ethnically and religiously inclusive insurgents. The latter aspect has made it difficult to consider the former as resulting from the often-referred ethnic and religious divides in the Sudan. The simulation results show that these two aspects of the conflict can emerge even in the virtual and simplified environment (see Fig.5.2), and their analyses suggest that the two are indeed inseparable. One of the important implications is that both of the inclusive and the exclusive symbols upheld by the two antagonists have some "fitness" (i.e. difficult to uproot) in their respective localities, leading to the political, as well as military, deadlock in the country.

6 Managing Interdependence: Absolute or Relative Gains?

Different recognition of national interests can result in different behavior, and thus in different international order. The well-known rivalry between neo-liberalists and neo-realists is examined by Suzuki (2003, 2004) in the case of the linkage of trade negotiation and international trade regime as follows.

6.1 Theoretical Background

If concerned with absolute gains alone, the state is interested only in its own wealth, and hence evaluates trade agreements regardless of other states' gain. Based on this assumption, neo-liberals (and orthodox economists) assert that creation and maintenance of an open world trade system is possible, with the help of the repetition of interaction and/or configurations of global wealth (Keohane 1984).

Proponents of relative gains, mostly political realists, claim that cooperation among states is not so easy. They insist that the most important objective of a state is to survive, and that power, which matters in relation to others and thus is a relative concept, counts for its behavior. States often refuse to make an agreement that favors the opponents more (Grieco 1988, 1990). Therefore, the resulting trade system is expected to be less open.

Then, if virtual states with absolute/relative gains are engaged in trade negotiation with one another, do they create orders predicted by these theoretical considerations? Which assumption generates the result closer to the reality?

6.2 Overview of the Model

Each state is concerned with its trade dependency, bargaining power, and optimal trade dependency from its economic size and current amount of trade. The state calculates them based on well-accepted premises: smaller nations tend to favor higher trade dependency; small and dependent nations have less bargaining power; profit from trade enlarges the size of economy, etc. In the next phase, every state randomly selects a negotiating partner, and begins negotiation to increase/decrease the amount of trade between the two. If preferences of the two coincide, they agree. If not, the stronger will prevail. However, the state pursuing its relative gain avoids agreements if the counterpart gets too much in comparison with own. Although the negotiations are bilateral, each state usually carries out trade multilaterally (**Fig. 6.1**).

6.3 Simulation Results and Implications

The result is surprising. Contrary to intuitive or theoretical prediction, trade dependency is the higher when states maximize relative gains than otherwise. The same result is observed even when the contribution of trade to economic growth is set at zero. Moreover, not only the ratio of trade but also the total real amount of trade was larger.

Fig. 6.2 shows the resulting patterns. In a world where states pursue absolute gain (**left**), all states line up on the most prefered trade dependence. The larger is the state's market, the lower the dependence. This is what states want, and all economies increase in size although smaller states tend to grow more rapidly. When states pursue relative gain (**right**), trade dependence of large states are almost same as those in absolute one. However, the small states' dependence tends to deviate greatly from the optimal points, in both upper and lower directions, and their economic growth is hamperd by excessive or insufficient trade. The only difference between the two cases is that states pursuing relative gain do not agree to certain deals that those pursuing absolute gain agree. Small difference of operational rules seem to create very different orders. This result implies the need to test theories and premises in the context of interplay of agents. Though the empirical figure is not shown, the right figure is similar, which suggests that national leaders are more concerned with relative gain.

7 Towards a User-Friendly, Multi-Purpose MAS Platform

Believe or not, all of the above four examples of MAS-based studies employ the same software package for MAS. This simulator is called "KK-MAS" that was originally developed by the Kozo Keikaku Engineering (KKE) some years ago. Now, with the close collaboration of the KKE, we have been developing "JAVA KK-MAS" (tentative name). Since April 2005, a so-called JAVA KK-MAS has been under the field test.

7.1 Need for a User-Friendly, Multi-Purpose MAS Platform

There exist some tools for MAS, but some difficulties are found to use them. For example, Starlogo is a very user-friendly simulator, but is not suitable to carry out academic investigations. On the other hand, Swarm can be used for developing complicated models, but it requires knowledge of the Objective-C language, which is a high barrier for most of social scientists. Repast, which was developed at the University of Chicago and Argonne National Laboratory, is now prevailing. Truly it is easier than Swarm, but it still requires the knowledge of the Java language. This object-oriented language is also difficult for social scientists to learn.

In order to overcome those linguistic and mathematical barriers, KK-MAS was developed so user-friendly as to meet potential demand of social scientists. KK-MAS requires minimal knowledge of programming languages so that those who understand the BASIC can develop complicated models that are suitable for social sciences.

7.2 The Concept of a New Platform

JAVA KK-MAS was designed to realize four objectives simultaneously: (1) to make model building as easy as possible, (2) to enable to construct as many types of models as possible, (3) to make analytical power as high as possible, and (4) to operate on various types of computers. While the original KK-MAS operates only on the Microsoft Windows OS, JAVA KK-MAS is programmed by the Java language, and therefore not dependent on the OS. The user does not have to know of the Java language. The only thing the user has to do is to build models simply by writing some "rules" in a BASIC-like manner.

As for the model, the generic structure is hierarchical ("universe"-"space"-"agent"), and the user is asked to create one or more spaces and one or more agents of one or more agent-types. Various functions are provided to support model building. The simulator controls and carries out all the process of simulation according to the model built. On JAVA KK-MAS, simulation can be executed, displayed and saved under various optional conditions that the user can choose.

JAVA KK-MAS has various advantages over Windows KK-MAS in addition to the fact that it can be installed on various OS platforms. In general, it has become more user-friendly, more flexible and more powerful. Some of the major improvements that have been implemented thus far include (1) a more strictly controlled scheduling of executions, (2) the definition of space and direction consistent with ordinary mathematical expressions, and (3) various functions for debugging. Those who are associated with academic institution, such as professors, graduate students and undergraduates, can participate in the on-going field test.

Readers are encouraged to join (http://www.kke.co.jp/iit/mas/index.html).

8 Possibilities of MAS

MAS has been introduced to social science in recent years, and many directions for MAS-related studies are yet to be explored. With new innovations in method and technique, novel uses of MAS may be invented.

8.1 From Multi-Agent Simulation to Mega-Agent Simulation

Although information technology is developing very fast, and the computer performance is becoming faster and smarter, it takes impracticably long time to conduct MAS as the number of agent increases. There are many MAS's in which tens of agents or hundreds interact each other; there are some MAS's that make thousands of agents interact. In order to simulate a cell automaton of 100x100, one has to build a model of ten thousand agents, and the execution of simulation may take hours depending on the complexity of interactions. In fact, there are few MAS's in which tens or hundreds of thousand agents interact. An exceptional MAS was carried out by Lomborg (1996) where more than a million agents interact more than 300 thousand times in a single run.

Once Mega-agent simulation, i.e., MAS involving millions of agents instead of thousands, becomes practical to conduct, the analysis of artificial society will change qualitatively, not to mention quantitatively. A big city or a small nation-state can be represented within the computer. Such technology is already used in analyzing the formation of a galaxy, or forecasting the weather next day or the climate next decade. A user-friendly mega-agent simulator may be developed for social scientists in near future.

8.2 Hybridization: Human-Agents Play with Artificial-Agents

Another possibility would be the hybridization of MAS, in which human players interact with programmed agents. Despite that the original KK-MAS or JAVA KK-MAS is not developed especially to build hybridized models, hybridization is possible. Hybridized models can be in fact built on the simulator in hand. In fact, a hybridized MAS model of the PD game has been built by Suzuki (forthcoming). This model creates a world of computer and human agents who are engaged with each other in the iterated PD game in a manner similar to Axelrod's well-known PD-strategy contest (1984). All agents, including one human player, are paired and play the game.

In this particular model shown in **Fig. 8.1**, seven computer agents identified as 00 through 06 and a human player (identified as 07) make 4 pairs, and each individual makes 50 choices before the pairs are changed randomly. In the default setting, the pay-off for mutual cooperation is 1 point, and 0 point for mutual defection. If one player defects while the other cooperates, the former gets 2 points and the latter loses 1. At the outset, each computer agent chooses one strategy from 10 famous strategies, such as All-C, All-D, Tit-For-Tat, Friedman, Tullock etc. In this case, "0" as indicated, the strategy is randomly chosen for all computer agents. Three small windows in **Fig. 8.1** show the latest result for the human player, total scores, and current pairs. On the right side there is the control panel for inputting the human player's hand, and for designating computer agents' strategies.

The model is so flexible that the model-builder (game-supervisor) can change the gaming environments. The human player may or may not know the strategies of the other agents. It is easy to change the gaming conditions: to randomize the number of iteration, to specify, change or randomize the strategy of computer agents, to change the score of pay-off matrix, etc.

Doubtlessly, this model (and probably hybridized MAS in general) is suitable for educational purposes. By participating in the simulation, one can experience and better understand the situation. For example, understanding the nature of the PD and possible strategies from a textbook may be challenging task for a kid, but if he/she is thrown into virtual jail and face the choice, the situation would be immediately clear to him/her, which can be called a "serious game". Alternatively, high school students may learn the importance and the difficulty of cooperation in global environment problems.

As for scholarly purposes, hybridized MAS may be useful in analyzing the nature of human decisions. Testing human players in anonymous or identifiable PD games to see how he/she reacts would be one example. Discovering a more human-like strategy would be another possibility. Just as we learn from computer simulations, so do the computer agents from us. If we use genetic algorithms, computer agent would learn strategies from human players, and this in turn might be utilized for better understanding of human behavior.

8.3 Networking: Agents Traverse Different "Universes"

Another innovative idea is the construction of various "universes" within a computer or on different computers connected one another. If MAS is conducted in each of "universes" independently, it is nothing but simultaneous runs of MAS. If different computers share a task, it may be conventional grid or parallel computing. Innovation exists where the same agent can move around on different "universes." The rules and/or roles that are applicable to agents can differ from one "universe" to another. Thus, behavior of agents depends on the "universe" in which they "live."

This innovation will make possible various types of MAS. For instance, based on a minimal common assumption, model-builders can construct their own model as they wish, and agents may behave differently depending on where they are in accordance with the model that applies to them. A particular agent who gave a lecture at a university campus in a model, may become a commuter who drives a car going back home. At home, the same agent may play roles of mother and wife. The same agent act differently on the situations which are constructed differently. In a word, different situations can be modeled on different computers by different architects. "Network MAS" may broaden not only the model specification but also the way of use drastically.

8.4 Concluding Remarks

There are fertile fields of political science for MAS. More political scientists will hopefully become interested in MAS-based analyses of political phenomena, and they will use MAS in their respective research topics. With user-friendly software for MAS, this new analytical method would be used oftener and more extensively, and would be recognized as useful. So-called "JAVA KK-MAS" is being developed so as to be one of such tools. This and other simulators will provide political scientists with a new horizon of research.

References

Allison, G.: Essence of Decision. Little, Brown, Boston (1971)
Allison, G. and P. Zelikow: Essence of Decision, 2nd ed. Longman, New York (1999)
Axelrod, R.M.: The Evolution of Cooperation. Basic Books, New York (1984)
Bhavnani, Ravi: Adaptive Agents, Political Institutions and Civic Traditions in Modern Italy, JASSS 6(4) (2003)
Bissey, Marie-Edith Mauro Carini and Guido Ortona: ALEX3: a Simulation Program to Compare Electoral Systems, JASSS 7(3) (2004)
Bremer, Stuart A. and Michael Mihalka: Machiavelli in Machina: Or Politics among Hexagons, in Deutsch, Karl W., Bruno Fritsch, Helio Jaguaribe and Andrei S. Markovits (eds.), Problems of World Modeling. Cambridge, Ballinger (1977).
Brouw, R.: Social Psychology. Free Press, New York (1965)
Cederman, L.E.: Emergent Actors in World Politics: How States and Nations Develop and Dissolve. Princeton University Press, New Jersey, London (1997)
Danno, E. and A. Tanaka: Kokusai System no Anteisei (Stability of International System), In Y. Yamamoto and A. Tanaka (eds.) Senso to Kokusai System (War and International System). Univ. of Tokyo Press, Tokyo(1992)
Downs, A.: An Economic Theory of Democracy. Harper and Row, New York (1957)
Duverger, M.: Les partis politiques: l'organisation des parties. Cours de droit, Paris (1954)
Eidelson, Benjamin M. and Ian Lustick: VIR-POX: An Agent-Based Analysis of Smallpox Preparedness and Response Policy, JASSS 7(3) (2004)
Epstein J.M. and Axtell R.: Growing Artificial Societies. MIT Press (1996)
Gatherer, Derek: Identifying Cases of Social Contagion Using Memetic Isolation, JASSS 5(4) (2002)
Grieco, J.M.: Anarchy and the Limits of Cooperation, International Organization 42. (1988) 485-507
Grieco, J.M.: Cooperation among Nations. Cornell University Press, Ithaca (1990)
Hishiyama, R.: Kokusai-Tsusin Rejime to Kosho (International Telecommunication Regime and International Negotiation), in Yamakage and Hattori (2002)
Johnson, Paul E.: Simulation Modeling in Political Science, American Behavioral Scientist 42(10) (1999).
Keohane, R.O.: After Hegemony. Princeton University Press, New Jersey, London (1984)
Kimura, K. and H. Oda: Kokusai-Kankyo Regime no Keisei (The Formation of International Environmental Regime), in Yamakage and Hattori (2002)
Kottonau, Johannes and Claudia Pahl-Wostl: Simulating political attitudes and voting behavior, JASSS 7(4) (2004)
Mitsutsuji, K.: Kokumin no Togo to Bunretsu (Integration and Disintegration of a Nation), in Yamakage and Hattori (2002).
Mitsutsuji, K.: The Artificial Political Society: Voters, Politicians and Parties in silico, Paper presented at 34th Annual Conference of International Simulation and Gaming Association, August 25-29, Kisarazu, Japan. (2003), also Working Paper No. 2 (our project)(2004)
Myers, D. G., and H. Lamm: The Group Polarization Phenomenon, Psychological Bulletin, 83 (1976), 602-627,
Nowak, A. et al., From Private Attitude to Public Opinion, Psychological Review, 97 (1990) 262-376.
Olson, B. D. and J. Suls: Self-, Other, and Ideal- Judgments of Risk and Caution as a Function of the Five-Factor Model of Personality, Personality and Individual Differences, 28. (2000)

425-436
Raczynski, Stanislaw: Simulation of The Dynamic Interactions Between Terror and Anti-Terror Organizational Structures, JASSS 7(2) (2004)
Sakamoto, T.: Kaso-Kokka niokeru Naisen no Saigen (Regenerating Civil War in a Virtual State), Working Paper No. 4 (our project)
Sakamoto T.: Funso to Doin (Conflic and Mobilization), Kokusaiseiji (International Relations) 140 (2005)
Sakamoto, T. and H. Hoshiro: Seisakukettei-Katei Simulation (The Simulation of Policy Formation Process), Working Paper No. 9 (our project) (2005)
Sakamoto, T. and H. Hoshiro: Seisakukettei-Katei Simulation II(The Simulation of Policy Formation Process II), Working Paper No. 12 (our project) (2006)
Sartori, G.: Parties and Party Systems. Cambridge University Press, Cambridge (1976)
Shimizu, T.: Shinka no simulation (Simulating Evolution), in N. Takahashi (ed.) Mirai Keisha Genri(the Principle of "Lean on Future"). Hakuto Shobou, Tokyo (1996), 29-54.
Srbljinovic, Armano, Drazen Penzar, Petra Rodik, and Kruno Kardov: An Agent-Based Model of Ethnic Mobilisation, JASSS 6(1) (2003)
Stoner, J. A. F.: Risky and Cautious Shifts in Group Decisions, Journal of Experimental Social Psychology, 4, (1968) 442-459
Suzuki, K.: Recognition of Interest and International Trade Negotiation. Paper presented at 34th Annual Conference of International Simulation and Gaming Association, August 25-29, Kisarazu, Japan. (2003), also Working Paper No. 1 (our project)(2004)
Suzuki, K.: Hybridized Prisoner's Dilemma Game, Working Paper (our project) (forthcoming)
Yasuda, S: GA niyoru Shujin no Dilemma Jikken (Experiments of Prisoner's Dilemma employing GA). Matsuyama Daigaku Ronso 13(2,3,4) (2001)
Yamakage, S. and S. Hattori (eds.): Computer no nakano Jinko Shakai (Artificial Society inside the Computer). Kyoritsu Press, Tokyo (2002)
Yamamoto, K.: Shakai-keiyaku no Ronri to Kiketsu (The Logic and Outcome of Social Contract), in Yamakage and Hattori(2002)
Yamamoto, K.: Kokusai-seijigaku no Simulation (Simulation in International Relations), Toyo Bunka Kiyo, 144 (2003)
Yamamoto, K.: Nation no Fukuzatsusei (The Complexity of Nations), Ph.D. dissertation (Univ. of Tokyo) (2004)
Yamamoto, K., M. Tamada, and A. Tanaka: Senso to Domei no Kokusaishakai (International Society of War ane Alliance), in Yamakage and Hattori(2002)

Tables and Figures

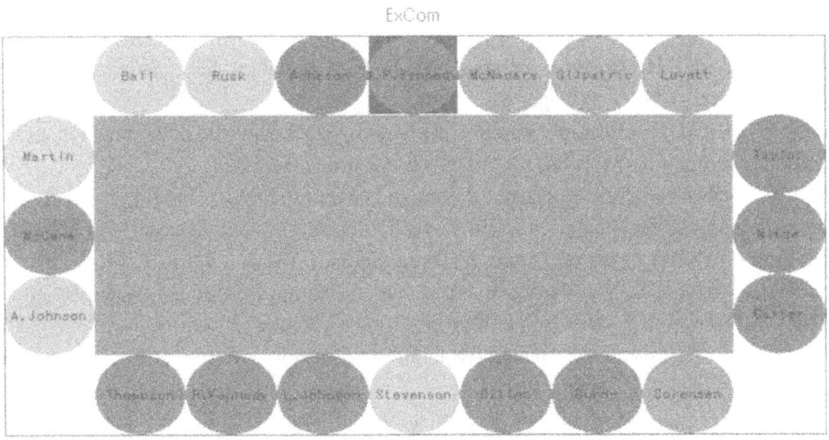

Fig. 3.1. The ExCom. Not all members continuously participated in weeklong discussions.

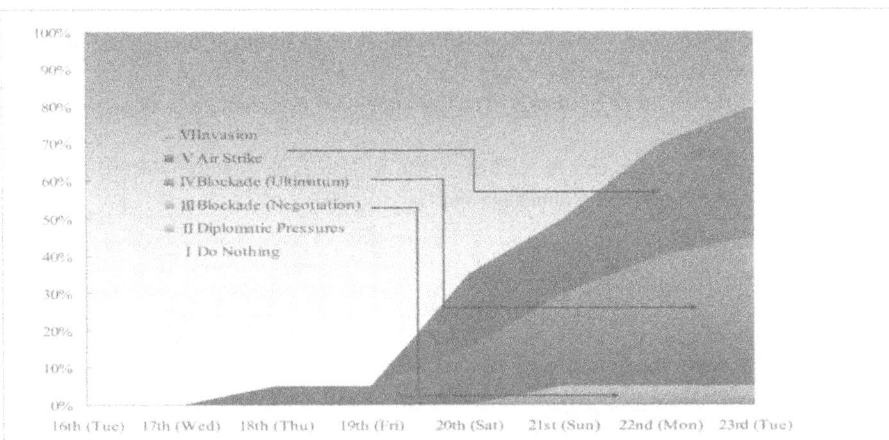

Fig. 3.2. Majority opinion formation: change of the frequency distribution of majority opinions. This is the result of twenty trials.

Table.4.1. Average Number of Political Parties on Various Conditions

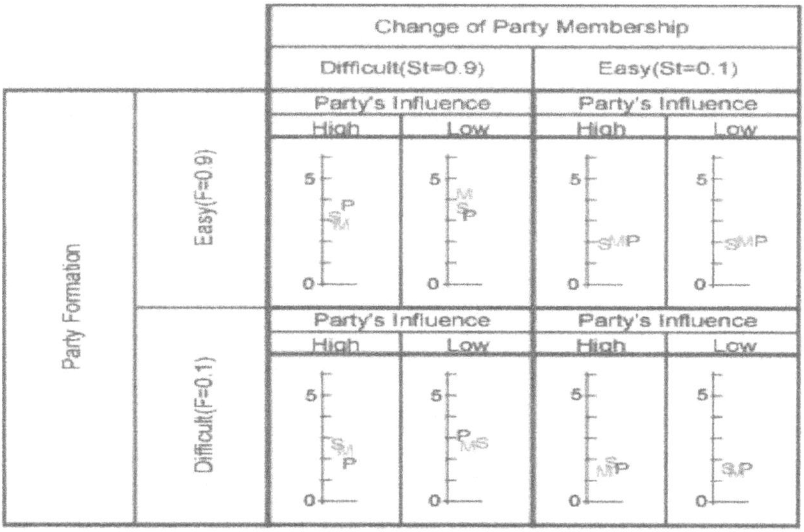

* Voter 1000, Seats 15, 10 times Elections
* S = single seated electoral district system (15 seats/ 15districts)
* M = medium district system (15 seats/ 5 districts)
* P = one didtrict system (15 seats/ 1 district)

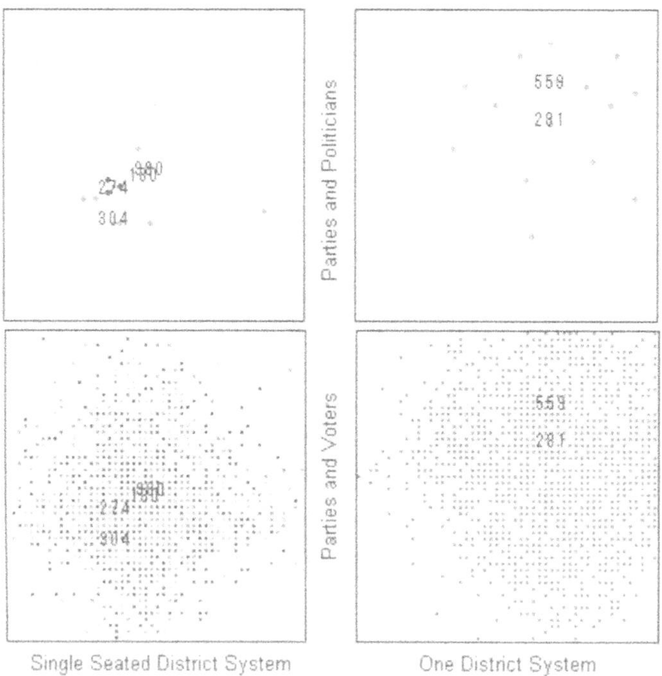

Fig. 4.1. Distribution of Voters, Politicians and Parties under the Single Seated District System and the One District System. Numerals denote political parties' ID.

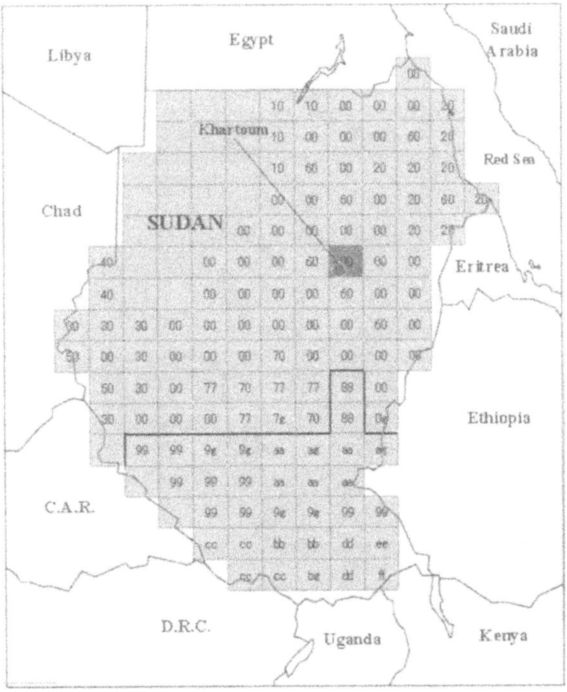

Fig. 5.1. The Virtual Sudan. A 2-digit string on each cell represents ethnicity and religion of the corresponding agent ("community"). For example, a "9g" community is ethnically Dinka ("9") and religiously Christian ("g"). Empty cells denote uninhabited areas.

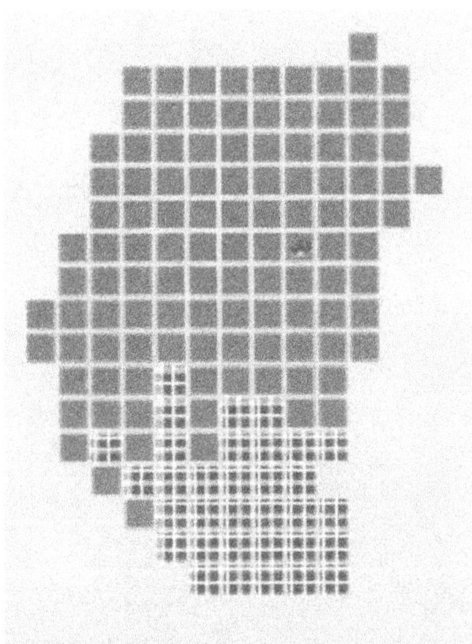

Fig. 5.2. Territorial division of the virtual Sudan (a typical case)

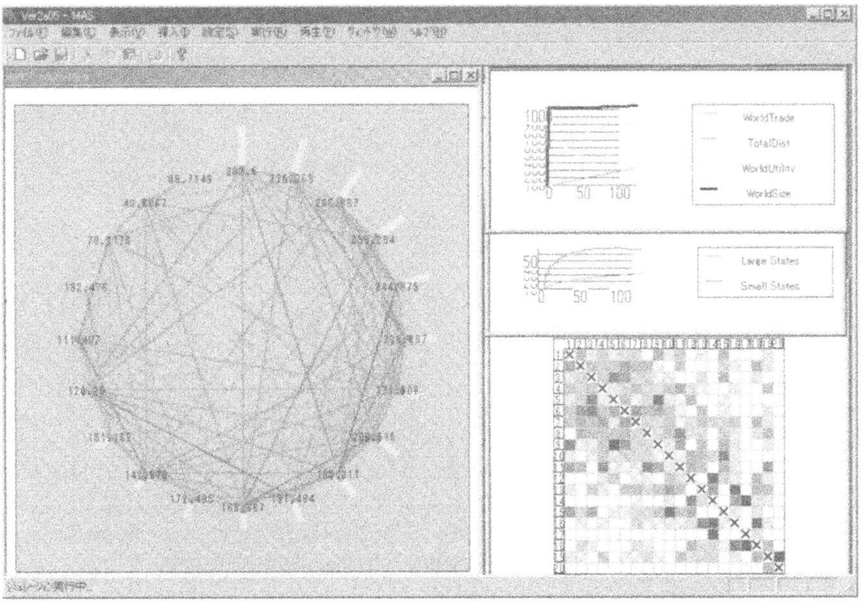

Fig. 6.1 The Trade Negotiation Model. The bar on each node shows its size and trade dependency, and the thickness of the lines between them indicates the amounts of trade. Diagram on the lower right shows each state's dependency on the others.

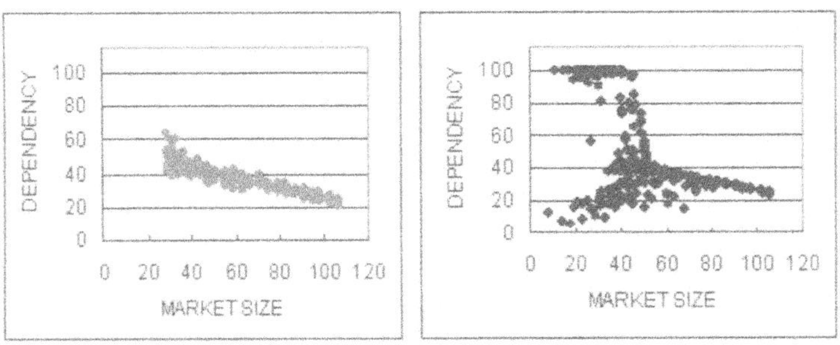

Fig. 6.2 Market size and dependency. States with absolute gains (left), relative gains (right). The simulation is run 50 times and each run contains 200 rounds of negotiations.

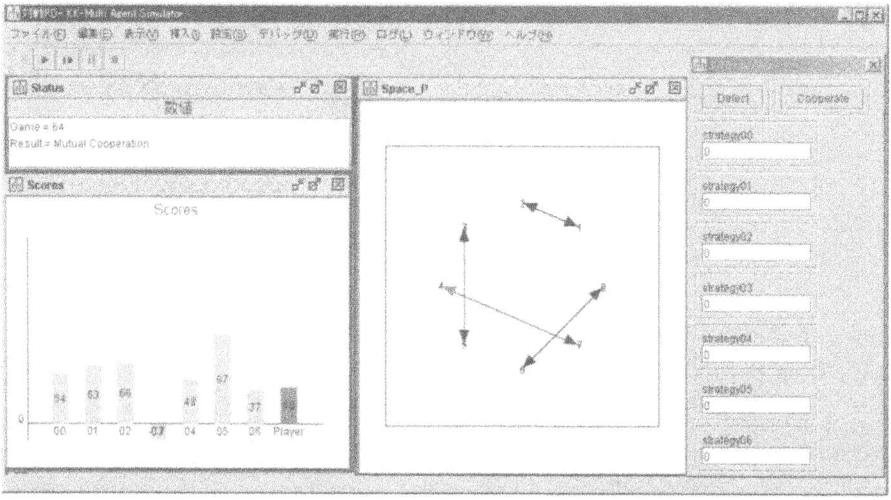

Fig. 8.1. Hybridized Prisoner's Dilemma Game. The human player chooses his/her hands by clicking the button "Cooperate" or "Defect" on the control panel (upper-right).

Replication of Sugarscape Using MASON[1]

Anthony Bigbee, Claudio Cioffi-Revilla, Sean Luke

Center for Social Complexity and Evolutionary Computation Laboratory
George Mason University, Fairfax, VA 22030 USA ccioffi@gmu.edu

1 Introduction

The purpose of this research was to replicate the Sugarscape model (Eptstein and Axtell 1996) and simulation outcomes as described in *Growing Artificial Societies* (*GAS*). Sugarscape is a classic agent-based model and contemporary simulation toolkits usually only have a very simple replication of a few core rules. There is scant evidence of significant replication of the rules and simulation outcomes; code supplied with Repast, Swarm, and NetLogo implement a minority of the rules in Sugarscape. In particular, the standard Repast distribution only implements **Growback**, **Movement**, and **Replacement**. Sugarscape implementations in these toolkits are clearly provided only as basic demonstrations of how well-known social models might be implemented, rather than complete achievements of scientific replication.

A major goal included assessing the maturity of the new MASON toolkit to replicate Sugarscape. MASON (Multiagent Simulator of Neighborhoods) "is a fast discrete-event multiagent simulation library core in Java, designed to be the foundation for large custom-purpose Java simulations, and also to provide more than enough functionality for many light-

---

[1] Presented at The 4th International Workshop on Agent-based Approaches in Economic and Social Complex Systems (AESCS 2005), Pacific-Asian Association for Agent-based Approach in Social Sciences (PAAAA), Tokyo Institute of Technology, Tokyo, July 9–13, 2005. The authors thank Joshua Epstein and Robert Axtell for comments on Bigbee (2005).

weight simulation needs." (Luke et al. 2005). Since MASON was designed to be a tool for social science research, among other uses, replication of one of the most recognized agent-based social science models would demonstrate its maturity and usability for its intended purpose.

Replication of well-known models is also important given the relative novelty of agent-based modeling in social science. Better tools and technique for lowering barriers to entry by social scientists are desirable outcomes.

2 Approach

Epstein and Axtell (1996) offer a framework – Sugarscape – for agent-based modeling and simulation that revolves around the following elements: agents, environment, rules. Epstein and Axtell state that the defining feature of the Sugarscape/artificial society model is that "fundamental social structures and group behaviors emerge from the interaction of individual agents operating on artificial environments under rules that place only bounded demands on each agent's information and computational capacity." Computationally, Sugarscape rests on an 'object-oriented' approach consisting of:

Instance variables representing agents' internal states or attributes (such as sex, age, wealth);

Methods for agents' rules of behavior (such as eating, trading, combat);

Encapsulation of agents internal states and rules to facilitate agent-based model construction.

Details regarding object-oriented (OO) techniques in Sugarscape are generally omitted from *GAS*. Appendix A contains a short section on OO techniques used and considered. Polymorphism is not discussed and inheritance was considered but was not used due to "efficiency considerations .... In total, each agent has over 100 methods." By comparison, the single agent class in MASON Sugarscape has approximately 32 methods, although only 75-80% of all Sugarscape rules were implemented. A prototype implementation of Sugarscape using ASCAPE appears to employ polymorphism and inheritance.

2.1 MASON

Although the MASON distribution includes a variety of graphics, utility, and other supporting infrastructure and examples, discussion of MASON in this paper focuses on timing and scheduling. Controlling and determining what behaviors are executed and when are critical aspects of simulation. In addition to overt requirements – such as an agent having to scan its surroundings and move to a site before it can harvest resources – synchronous and asynchronous interactions occur between entities. A necessary, but not sufficient, requirement is that actions are executed in time exactly as desired—this requires a precise scheduling mechanism.

2.1.1 Scheduling and Time

The MASON Version 10 Schedule class sequences and executes objects that implement the `Steppable` interface. Classes implementing `Steppable` have a `Step` method for the schedule to initiate behavior/execution. A primary construct in `Schedule` is order, a collection of executable objects that enable a deterministic sequence of execution. For each time step, all order zero objects are executed first, followed by order one, and so forth. Implementation of rules in specific orders is a key aspect of this replication and orders for agent and environment rules are currently specified in the Sugarscape class as constants. These constants could easily be redesigned as parameters in the primary configuration file for more flexible experimentation.

`Seasons`, in addition to statistics, charts and logging, uses another MASON class – `MultiStep` to enabling `Steppables` – to be executed less than once every time step but with a regular periodicity. `Schedule.setRepeating` method that has an interval has better performance than `Multistep` when orders have multiple `Steppables`

Within each order, the `Steppable` entities are executed once in a random sequence, and the sequence is randomized every time step. The other critical timing mechanism used in this implementation is that agent instances themselves do not determine which of their rules are executed, nor in what sequence. Instead, a member instance named `RulesSequence` wrap `Sequence`, a MASON class for holding a static sequence of `Steppables`. This static sequence contains a collection of rules, each of which is invoked in turn. A benefit of using `Sequence` is that the `step()` method in the agents or environment objects are small and simple, primarily calling `step()` for its `RuleSequence`. The `RuleSequence` instance, in turn, calls each rule in the original order specified when the rules were added during initialization. `Sequence` only has order *m* rules time complexity as opposed to order *n* entities x *m* rules complexity in direct schedule usage .

To provide for a flexible experimentation, a primary configuration file specifies rule execution order during runtime. Using *reflection* facilities of the Java language, initialization code translates text names for rules classes into object instantiations. This allows quick way to specify desired rule sequences without source code recompilation.

Finally, there are rules that involve cellular automata (CA) type synchronous treatment of all instances of a class (i.e. all entities of one type). **Pollution Diffusion** (**D**) is an environment rule in which the states of all sites synchronously update without using MASON scheduling machinery.

2.2 Space-time Interactions

Many agent-based models have discrete space aspects that UML sequence diagrams cannot visualize. As an example, interactions in space and time are illustrated in Figure 1. This diagram provides insight into the local effects of agent harvesting when a) sites grow back during the harvesting time step, and b) sites are restricted from growing back until one full time step after the time step during which harvest occurred. The blue circle depicts the location of the agent at each time step. Bigbee (2005) has an extended discussion on this phenomenon and emergent behavior.

2.3 Rules Implemented

Table 1 describes rules implemented from the classic model and whether the rule can be added, removed, or reordered simply by editing the `agent_rules_sequence` or `environment_rules_sequence` lines in the runtime configuration file.

Total source lines of code are approximately 3500 as counted by the SLOCCOUNT tool (Wheeler 2005). Total source lines of code for the rules is 934, with the ratio of rules code to other code being approximately 1:3. The non-rules source lines of code involve model initializing, graphs and statistics, logging, and parameter sweeping. The total SLOC count for MASON version 8 itself, not including the supplied demonstration applications, is approximately 18000. This includes many classes not used by MASON Sugarscape, including other portrayals and 3-D visualization infrastructure.

3 Results

Table 2 documents simulation outcomes and rule sets investigated as part of replication. Qualitative and other criteria are described in the table, such as whether aggregate statistics and shapes of graphs were matched, as well as an overall level of replication achieved—exact, general, or partial. The overall pattern of replication outcomes is that the outcomes documented in *GAS* were generally replicated in MASON Sugarscape simulations. The most successful replications occurred for outcomes that did not involve movement/welfare-dependent agent survival; these outcomes included **Culture**, **Pollution Diffusion**, **Seasons**, and other spatial phenomena. Bigbee (2005) provides detailed discussion of each outcome and issues in replicating.

4 Summary

The research yielded partial replication of the Sugarscape outcomes described in GAS. A major lesson from this research is the difficulty in understand and constructing simulation models that appear simple yet have complex emergent behavior.

Software engineering is a young field, although tools and techniques have emerged to support error/bug reduction, automated testing, and faster development. While OO-based software has been touted as an effective way of constructing software applications, OO methodologies do not eliminate cognitive error nor complexity in development. Close examination of the psychology of software development is beyond the scope of this thesis, but attention to the biases and heuristics literature and cognitive errors literature might yield some techniques and suggest tools appropriate for effective development of agent-based models.

Bigbee (2005) offers a list of lessons learned and 10 heuristics and ideas to promote successful replication. Social scientists creating or replicating agent-based models face a variety of challenges encountered in other fields and much work remains to be done to lower barriers to good science in this field.

References

Bigbee A (2005) Replication of Sugarscape Using MASON. Unpublished master's thesis, George Mason University, Fairfax, VA.

Densmore O (2005) Sugarscape. Retrieved April 1, 2005, from http://backspaces.net/Models/sugarscape.html

Doran J (2000) Questions in the Methodology of Artificial Societies. In. Suleiman R, Troitzsch K, Gilbert N (eds), *Tools and Techniques for Social Science Simulation.* : Physica-Verlag, Heidelberg

Epstein J, Axtell R (1996) *Growing Artificial Societies: Social Science from the Bottom Up*. Brookings Institution Press, Washington, D.C.

Hegselmann, R, Flache A (1998, June) Understanding Complex Social Dynamics: A Plea for Cellular Automata Based Modeling. *J Artificial Societies and Social Simulation* 3, Retrieved August 1, 2004, from http://www.soc.surrey.ac.uk/JASSS/1/3/1.html

Huberman B, Hogg T (1988) The Behavior of Computational Ecologies. In B. Huberman (Ed.), *The Ecology of Computation.* Amsterdam: North-Holland.

Kleiber C, Kotz S (2003) *Statistical Size Distributions in Economics and Actuarial Sciences.* Wiley, Hoboken, NJ

Kliemt H (1996) Simulation and Rational Practice. In R. Hegselmann, U. Mueller (Eds.), *Modelling and simulation in the social sciences from a philosophy of science point of view.* Kluwer, Dordrecht

Luke S, Cioffi-Revilla C, Panait L, Sullivan K, Balan G (2005) MASON: A Multi-Agent Simulation Environment. *Simulation* 81: 517-527

Nowak A, Lewenstein M (1996) Modeling social change with cellular automata. In R. Hegselmann, U. Mueller (Eds.), *Modelling and simulation in the social sciences from a philosophy of science point of view.* Kluwer, Dordrecht

Wheeler D SLOCCount. Retrieved April 1, 2005, from http://www.dwheeler.com/sloccount/

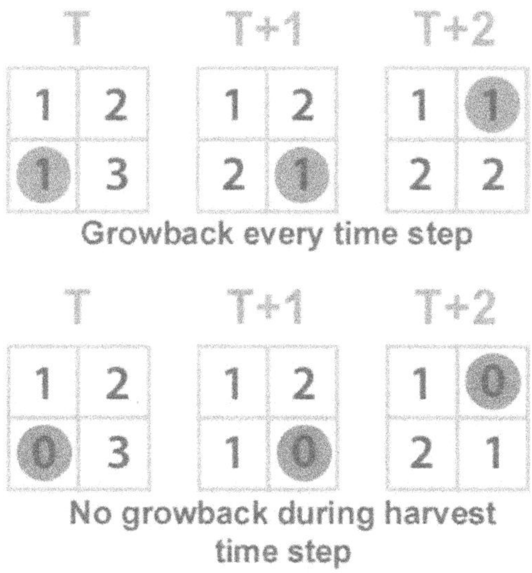

Fig. 1. Space time interactions under two Growback rules

Table 1. Implemented Rules

Symbol	Name	SLOC
G_α	Sugarscape growback	42
M	Agent movement	323
$R_{[a,b]}$	Agent replacement	17
$S_{\alpha\beta\gamma}$	Seasonal growback	16
$P_{\pi,x}$	Pollution formation	18
D_α	Pollution diffusion	56
S	Agent mating	163
None	Agent cultural transmission	
None	Group membership	128
K	Agent culture	
T	Agent trade	177

Table 2. Replication Outcomes

Outcome	Rule Set	Replication Criteria	Replication Achieved
Animation II-2 (p.29)	$(\{G_1\}, \{M\})$	Hiving, peak clustering, terrace sticking	Exact
Figure II-5 (p. 31)	$(\{G_1\}, \{M\})$	Small positive slopes, equally spaced lines, visually estimated line coordinates	General
Animation II-3 (p. 34)	$(\{G_1\},\{M,R_{[60,100]}\})$	Pareto distribution, maximum wealth bin	General
Animation II-4 (p. 38)	$(\{G_1\},\{M,R_{[60,100]}\})$	Gini coefficient evolution	General
Animation II-6 (p. 43)	$(\{G_1\}, \{M\})$	Visual wave phenomenon	General
Animation II-7 (p. 46)	$(\{_{S1,8,50}),\{M\})$	Seasonal clustering	Exact
Animation II-8 (p. 49)	$(\{G_1,D_1\}, \{M,P_{11}\})$	Migration patterns	Partial
Figure III-1 (p. 58)	$(\{G_1\}, \{M,S\})$	Stable time series	General
Animation III-1 (p. 58)	$(\{G_1\}, \{M,S\})$	Approximate stationary age distribution	General
Figure III-2 (p. 63)	$(\{G_1\}, \{M,S\})$	Diverging Vision, Metabolism	General
Figure III-3 (p. 64)	$(\{G_1\}, \{M,S\})$	Small amplitude oscillations	General
Figure III-4 (p. 65)	$(\{G_1\}, \{M,S\})$	Large amplitude oscillations	Partial
Figure III-5 (p. 66)	$(\{G_1\}, \{M,S\})$	Severe population swings, extinction	General
Animation III-6 (p. 75)	$(\{G_1\}, \{M,K\})$	Homogenous population	Exact
Figure III-8 (p. 77)	$(\{G_1\}, \{M,K\})$	Time series extremes, random group convergence	Exact
Animation IV-1 (p. 100)	$(\{G_1\}, \{M\})$	Peak hopping, small population	General
Figure IV-4 (p. 110)	$(\{G_1\}, \{M,T\})$	Significant trade volumes over time	Partial

X-MAS: Validation tool based on meta-programming

Yutaka I. Leon Suematsu[1,2], Keiki Takadama[1,3], Katsunori Shimohara[1,2] and Osamu Katai[2]

[1] ATR International - Network Informatics Labs. 2-2-2 Hikari-dai, Seika-cho, Soraku-gun, Kyoto 619-0288, JAPAN. {yutaka, keiki, katsu}@atr.jp
[2] Graduate School of Informatics, Kyoto University, Yoshida-honmachi, Sakyo-ku, Kyoto 606-8501, JAPAN. Email: katai@i.kyoto-u.ac.jp
[3] Interdisciplinary Graduate School of Science and Engineering, Department of Computational Intelligence and Systems Science, Tokyo Institute of Technology, 4259 Nagatsuta-cho, Midori-ku, Yokohama, Kanagawa 226-8502, JAPAN

Summary. Validation is an important issue in Agent-Based Modeling (ABM). Unfortunately, although indispensable, it is not a common practice in the community. The main reason is that there are no established validation procedures. In our previous work, we proposed the *cross-element validation* as a validation process that consists of detecting, analyzing and comparing the model's macro-behavior under different variations of its composite elements. This process requires performing several simulations of the model with modifications in some of its algorithms. Therefore, it is indispensable the availability of some tools that provides: (1) easy model implementation, (2) flexibility for easy model's elements exchange, and (3) construction of efficient code for accelerating the simulations. In order to support these requirements, this paper proposes the X-MAS toolkit, which facilitates the implementation and cross-element validation of ABM models.

Key words: X-MAS, agent-based model, cross-element validation, meta-programming

1 Introduction

Nowadays, agent-based modeling (ABM) is considered a prominent paradigm in the study of complex social systems. However, the main difficulty to be faced is the lack of validation methodologies, restricting the reliability of ABM models and their results. A well-know approach is the so-called *docking* or *alignment of computational models* (Axtell et al. 1996), which is a validation process where two models that deal with the same phenomenon are compared, in order to determine whether the results from one computational model match the results of another model, by extending their interpretations. Even, docking is an exciting approach for validation, this process leads to the following questions, which make it rather awkward: (1) how to modify a model without missing the essence or features of the other model, (2) how to define the comparison standards to judge models equivalence, and (3) how to manage the lack of access to the source code and detailed documentation of other researchers' models, especially old ones.

To overcome the above-mentioned difficulties of docking, in our previous work we proposed the concept of *cross-element validation* (XEV) (Takadama et al. 2003) as a process that consists of performing the validation in only one model, comparing the results of the model by using substitutable elements that compose the model. For this process, it is indispensable the availability of some tools that provides: (1) easy model implementation, (2) flexibility for easy model's elements exchange, and (3) construction of efficient code for accelerating the simulations.

In order to solve these difficulties, our research focuses on the implementation of a toolkit for supporting XEV. This paper proposes the Cross-element validation for Multi-Agent-based Simulation (X-MAS), which is a toolkit developed in C++, with cutting edge techniques of meta-programming to support the XEV and facilitate the implementation of models.

This paper is organized as follows: In chapter 2 a brief description of XEV and its difficulties are presented. Chapter 3 provides a brief description of X-MAS and its architecture. Chapter 4 shows some usage and advantages of the tool. The discussions are provided in Chapter 5; and finally the conclusions and future work are presented in Chapter 6.

2 Cross-element validation and its difficulties

2.1 Cross-element validation

Cross-element validation (XEV) is a validation process that performs the validation *within* a model, as opposed to docking, which consists of validation *between* models. XEV consists of detecting, analyzing and comparing the model's macro-behavior under different variations of its composite elements. Elements are, for instance, how individual agents determine their behaviors, store knowledge, or performing learning. By understanding whether and how different agent implementations affect the model's overall behavior, the reliability of the ABM model is expected to increase.

2.2 Difficulties of cross-element validation

XEV requires performing several numbers of simulations of one model with variations of its composite elements. The number of simulations is exponentially increased as both the number of replacements of a certain element and the number of elements to evaluate are increased. Thus, efficient performance of each simulation is required. Another difficulty is the modification of the simulation programs, which is a cumbersome. Due to the fact that XEV, even simple, is a tedious task, it is indispensable the availability of tools that satisfy the following three requirements: (1) easy model implementation, (2) flexibility for easy model's elements exchange, and (3) construction of efficient code for accelerating the simulations.

2.3 Difficulties for tool implementations

In our previous work (Leon et al. 2003), a first attempt to develop a tool for supporting the XEV was tested. The object-oriented programming (OOP) was employed because the notion of object suitably fits the agent concept. Like many available agent-based simulators, this attempt made extensive use of the inheritance and the polymorphism[1] potential of the OOP. For providing flexible libraries, they are implemented using common base types, defining the interface for user's own implementations. They allow defining functions or member functions that use base classes as parameters, receiving any type of descending classes, for generalizing the usage of the function. In the case of it is required to use some member functions of the descending class that are not described in the base class, a transformation from a base class to its original class (descending class) type is required. This is the so-called *cast operation*, and the original type must be explicitly described in the implemented function.

For the purpose of tools for XEV, the above-mentioned process is a drawback due to fact that it looses the benefit of type checking. In other words, careless use of functions that uses base classes as parameters may receive unexpected information producing a malfunction of the program. This kind of bug is very common. Additionally, excessive use of base classes makes it difficult to maintain the source code of the model. For instance, when explicit *cast operator* is required, the replacement of some algorithm of the model may require the implementation of the same behavior or algorithm again and again for each variation of the element. Also, updates on the algorithm require updates in all created variations. Finally, excessive use of polymorphism may affect the efficiency of the simulation, which is significant when implementing large-scale simulations.

In our first attempt for developing the tool, the above-mentioned difficulties were faced. Therefore, it could not fulfill the requirements; specifically, requirement 2 (flexibility for easy model's elements modifications).

3 X-MAS

3.1 Concept

The Cross-element validation for Multi-Agent-based Simulation (X-MAS) provides a rich framework that facilitates both the implementation and XEV of ABM models. Additionally, X-MAS facilitates *multi-intelligent* agent implementation

[1] Polymorphism is the ability to process objects differently, depending on their data type or class. More specifically, it is the ability to redefine methods for derived classes

by providing a framework that embeds the commonly used learning mechanisms and knowledge representation schemes for agents.

The proposed version of the X-MAS is a generic OOP libraries developed in Standard C++ with cutting edge techniques of meta-programming. It makes extensive use of STL[2] and BOOST libraries (Boost 2005). The use of C++ meta-programming (Abrahams and Gurtovoy 2004) makes it easier the exchange of elements of the model, satisfying the flexibility required for the XEV. X-MAS is portable because it can be compiled in several C++ compilers, regardless the operating system. Also, due to fact that X-MAS's libraries make extensive use of meta-programming techniques, it produces efficient code for good performance of the simulation. Thanks to these advantages, it was possible to develop the X-MAS that effectively fulfills all the requirements for XEV tools.

3.2 Meta-programming approach

X-MAS employs meta-programming techniques to provide well-designed generic libraries that allow a reusable mental model in approaching problems. It provides the same interface regardless of the platform used. Additionally, some pre-processing can be performed in compile time so that the code is faster. All of these features make X-MAS practical and a general-purpose ABM simulator.

The potential of this approach is that it provides a meta-language that allows the manipulation of types in compile time, allowing detection of errors regarding type check. In the example shown in listing 1, the logic of the agent is implemented only once (lines 1-12), but the evolution algorithm and genotype to be used are not explicitly defined. Only when the agent is instantiated the algorithm to be used is defined (line 15 and 16 for genetic algorithm and evolution strategies, respectively). It should be noted that the definition of the algorithm to be used is performed in compile time, evaluating the compatibility with the genotype.

Listing 1.Implementation of generic evolution algorithm-based agents

```
 1  template<class EvolutionAlgoType, class GenotypeType>
 2  class EA_Agent {
 3    EvolutionAlgoType evol_algo;
 4    GenotypeType[PopulationSize] Population;
 5    ...
 6    void evolve(GenotypeType* newPop) {
 7      ...
 8      evol_algo.nextGeneration(Population,newPop);
 9      ...
10    }
11  ..
12  };
13
14  // Instanciation of agents
15  EA_Agent <GeneticAlgorithm, GAGenotype> GAAgent1;
16  EA_Agent <EvolutionStrategy, ESGenotype> ESAgent1;
```

[2] STL means Standard Template Library

3.3 X-MAS architecture

X-MAS, as shown in figure 1, consists of three main libraries: (1) X-MAS core libraries, (2) Domain-specific libraries, and (3) Cross-element validation interfaces.

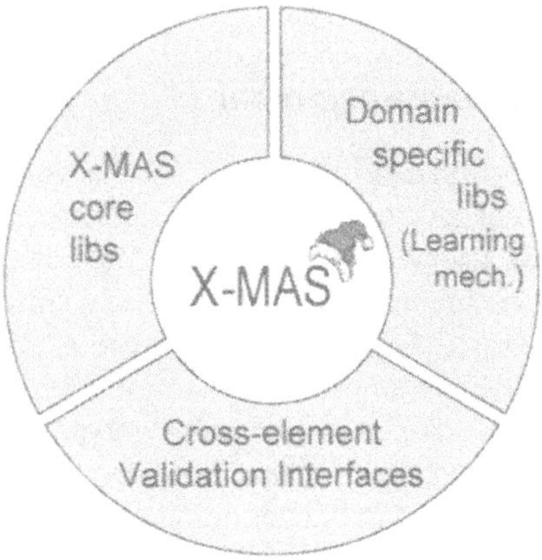

Fig. 1. X-MAS architecture

X-MAS core libraries

X-MAS provides a set of generic libraries and utilities for easy implementation of models. Several commonly used algorithms in model implementations are provided, which are highly customizable. Some implementations, for instance, consist of the scheduling of agent interaction, selection algorithm, and simulation control.

Domain-specific libraries

The X-MAS framework allows the implementation of domain-specific libraries by customizing some of the already available libraries. Intelligent agents' implementation is supported by providing rich libraries, including several learning mechanism implementations such as Reinforcement Learning (RL)(Sutton and Barton 1998), Genetic Algorithms (GA)(Holland et al. 1986) and Evolution Strategies (ES)(Back et al. 1991). It also provides genotype-based, rule-based, and other algorithms with discrete and continuous values as knowledge representations schemes. Each algorithm provides default settings and highly customizable frameworks to include variations of the algorithms. Parts of the algorithm can be replaced with user's algorithms, such as different crossover algorithms for GA.

Cross-element validation interfaces

As the main purpose of X-MAS, it provides a set of interfaces to interact with the model and to perform simulations with several variations of its elements. This module is in ongoing development. Due to the fact that there are not standard graphic libraries in the standard C++, which is the main disadvantage, some interfaces with Python and Java are planned for further development.

4 Usage

Customizing an evolution algorithm

X-MAS provides rich libraries for learning mechanisms. Assuming a model that consists on ES-based agents, as shown in listing 2, several variations of the algorithm can be assigned easily to agents (lines 7-15). For instance, *agent1* is created by using the default genotype and default algorithm (lines 7-8). By default, the algorithm will maximize the fitness of the genotype; if minimization is required, it can be expressed by adding *minimize* in the definition of the algorithm (line 10). The algorithm will work with users-defined genotypes. It only requires the definition of the *genotype_policy* that manages the user genotype. After assigning the policy (line 13), it is included in the definition of the algorithm (line 14). It should be noted that the selection of different elements of the algorithm itself is allowed.

Listing 2. Customization of an evolution strategies-based agent

```
 1  Template <class EvolAlgoType, class GenotypeType>
 2  class ESAgent {
 3  EvolAlgoType evol_algo;
 4  ...
 5  }
 6  // default algorithm
 7  typedef xmas:: evolution_strategy< > es_type1;
 8  ESAgent<es_type1, xmas::es::genotype_type<> > agent1;
 9  // default algorithm with minimization algorithm
10  typedef xmas:: evolution_strategy<default_,
    default_, default_, xmas::minimize > es_type2;
11  ESAgent<es_type2, xmas:: es_genotype<> > agent2;
12  // default algorithm with user_defined genotype
13  typedef xmas::genotype_policy<double, user_genotype,
    user_genotype_init>    genotype_policy1;
14  typedef xmas::evolution_strategy<genotype_policy1>
    es_type3;
15  ESAgent<es_type3, user_genotype > agent3;
```

5 Discussions

5.1 X-MAS as tool for validation of ABM models

XEV, that simulates a model with several variations of some its composite elements, can be easily performed with X-MAS. The X-MAS design provides a powerful tool not only for performing XEV, but also for performing sensitivity analysis of the model. The meta-programming approach and X-MAS allows direct implementation of the model's logic in a general way. Due to the fact that X-MAS makes some additional verification in compile time, the number of bugs is expected to be reduced. Unfortunately, the current version under development does not provide graphical interfaces to interact with the model in a friendly environment, but they are planned for the final release.

5.2 X-MAS as tool for implementation of ABM models

As it was observed in the example explained in section 4, X-MAS provides powerful libraries for easy implementation of ABM models. The provided generic libraries can be adapted for user implementations. This can be considered as a real reusable library, allowing inclusion of users modifications without manipulating the libraries itself. Therefore, maintenance of code is getting effortless. Although, in first term, extending the libraries of X-MAS will require knowledge of C++ and some experience with BOOST libraries, the return is precious.

5.3 X-MAS compared to some available tools

In order to help researchers to simulate their models, several ABM toolkits have been developed. Basically, they provide a set of libraries and frameworks to reduce the difficulties of the programming process, and to enhance the understanding of the outcomes (e.g., Repast (University of Chicago 2005), Swarm (Swarm Group 2005), Ascape (Parker 2001)). Those tools have been developed with OOP for providing flexible frameworks. Although efficient for implementing models, the XEV of models by using these toolkits is still a tedious task due to the fact that they do not fulfill the requirements of this validation process. The difficulties are the same as in our first attempt of X-MAS, described in section 2.3. In contrast, the current design of X-MAS was designed to fulfill the requirement for supporting the XEV. As a model implementation tool, its generic design allows the implementation of several types of models. Additionally, X-MAS is suitable for exploring complex social systems composed of large number of *intelligent agents*.

6 Conclusions

XEV is a process that consists of performing the validation in one model, comparing the results of the model under variations of its composite elements. For this process, it is indispensable the availability of tools that provides: (1) easy model implementation, (2) flexibility for easy model's elements exchange, and (3) construction of efficient code for accelerating the simulations. The Cross-element validation of Multi-Agent-based Simulations (X-MAS) was proposed, which satisfactory fulfills all requirements for XEV. X-MAS was developed in C++ with advanced techniques of OOP and meta-programming. It provides generic libraries that are customizable to user's necessities. The X-MAS not only supports XEV, but also supports sensitivity analysis of the model. Additionally, it is a powerful framework for ABM model implementations. Further research includes: (1) implementation of GUIs, (2) inclusion of logging libraries, (3) comparison of X-MAS with available tools and (4) performing XEV of several models.

Acknowledge

This research was conducted as part of the 'Research on Human Communication' with funding from the National Institute of Information and Communications Technology (NICT), Japan. This research also was supported in part by the Sasagawa Scientific Research Grant from the Japan Science Society. The research reported here was also supported in part by a Grant-in-Aid for Scientific Research (Young Scientists (B), No. 17700139) of Ministry of Education, Culture, Sports, Science and Technology (MEXT). We would like to express our thanks for all of the above support.

Reference

Abrahams D and Gurtovoy A (2004) *C++ Template Metaprogramming*, Addison Wesley Professional

Axtell R, Axelrod R, Epstein JM and Cohen M (1996), Aligning simulation models: A case study and results, *Computational and Mathematical Organization Theory (CMOT)* 1(1): 123-141

Back, T., Hoffmeister, F. and Schwefel, H., 1991, "A survey of evolution strategies", in *Proceedings of the Forth International Conference on Genetic Algorithms*, pp. 2-9.

Boost Libraries (2005) Boost Library Home Page, http://www.boost.org/

Holland J, Holyoak K, Nisbett R and Thagard P (1986) *Induction*, the MIT Press

Leon YI, Takadama K, Nawa E, Shimohara K and Katai O (2003) The X-MAS: Toward simulation systems for cross-model-validation of multiagent-based simulation", *Meeting the Challenge of Social Problems via Agent-based Simulation*, Springer-Verlag, pp 171-184

Parker M (2001) What is Ascape and Why Should you care?, Journal of Artificial Societies and Social Simulation 4(1)

Sutton R and Barto A (1998) *Reinforcement Learning: An Introduction*, the MIT Press

Swarm Development Group (2005) Swarm Home Page, http://www.swarm.org/

Takadama K, Leon YI, Sugimoto N, Nawa E and Shimohara K (2003) Cross-element validation in multiagent-based simulation: Switching learning mechanisms in agents, *Journal of Artificial Societies and Social Simulation* 6(4)

University of Chicago (2005) Repast Home Page, http://repast.sourceforge.net/

Hardware Implementation of Parallel SOARS using FPGA based Multiprocessor Architecture

Hideki Tanuma[1], Hiroshi Deguchi[2], Tetsuo Shimizu[1]

[1]Institute of Medical Science, University of Tokyo, 4-6-1 Shirokane-dai, Minato-ku, Tokyo 108-8639, Japan
[2]Department of Computational Intelligence and Systems Science, Tokyo Institute of Technology, 4259 Nagatsuta-cho, Midori-ku, Yokohama, Kanagawa 226-8503, Japan

Summary. We are developing an agent-based simulation language, SOARS (Spot Oriented Agent Role Simulator), aiming standard descriptive language to express complicated social system. SOARS is a simulation language with original model concept to describe agent's interaction. Currently, SOARS processor is implemented by JAVA, and agent rule is successively processed. However, model concept of SOARS is originally headed for parallel execution of agent rule, and it is preferable to prepare for high-speed implementation performing right parallel operation by common simulation description. Therefore, we decided to investigate the problem in language specification at the time of parallelizing, construct and monitor the system that is small-scaled but enables actual parallel operation.

Parallel processing with high granularity like the agent rule of SOARS is an interesting theme of computer science in itself. Such parallel processing is considered suitable to be implemented by multiprocessor core architecture CPU that is thought to come out one after another. The system available, however, is hard to obtain at present. Therefore, we developed unique processor architecture using FPGA (Field Programmable Gate Array) that recently shows remarkable progress and can be obtained at low cost.

Key Words. SOARS, Agent based simulation language, Complex systems, FPGA, Parallel processing, Multiprocessor architecture, Hardware implementation

1 Introduction

We often use simulation analysis in agent-based modeling. These days Ascape[1] by The Brookings Institution and Repast[2] by the University of Chicago and Argonne National Laboratory are becoming popular. In the history of agent-based modeling, Swarm by the Santa Fe research institute has been very influential. Repast is called the social science version of Swarm. In Japan, MAS (Multi-Agent Simulator)[3] has been developed under the influence of Swarm for educational purposes. Repast, Ascape and MAS are descendants of Swarm which has been developed nor for social sciences but for cellar-type simulation.

In this paper we introduce a new type of modeling framework for the agent-based dynamic system (ABDS) and develop an agent-based simulation language called SOARS (Spot Oriented Action Role Simulator).

SOARS has been developed for infection protection against SARS in hospitals. In hospitals, there are large numbers of agents: doctors, nurses, patients, radiographers, inspectors, office workers and volunteers, who move about among many spots.

2 Parallelizing of simulation language SOARS

(1) Overview of SOARS modeling concept

In SOARS we introduce several notions for modeling an agent-based dynamic system; they are the agent, the spot, the resolver, information and the physical object.

The spot is an abstract notion of a localized field or a place where agents interact with each other. Agents move among spots. Each spot and agent has a state that is described by equipped objects. Spots are used to represent not only a concrete physical place but also an abstract place for interaction such as a committee. In the following example we use concrete spots in a hospital, such as a consultation room or a hospital ward. A spot is equipped with a state object that is called an information object or a physical object. A spot is also equipped with a special type of object called a resolver. A resolver on a spot describes an interaction among agents.

In example of a hospital infection model, a group of agents consists of office staff, cashier staff, reception staff, nurse, doctor, examiner, visitor, and patient. A group of spots inside a hospital consists of locker room, office, medical office, reception, examination waiting area, examination room, nurses' station, consultation

[1] http://www.brook.edu/es/dynamics/models/ascape/default.htm

[2] http://repast.sourceforge.net/

[3] http://www2.kke.co.jp/mas_e/MASCommunity1.html

waiting area, consultation room, ward, cashier, and mortuary. Spots which exist outside a hospital are home, business office and school.

We use step as a basic time unit of a discrete time dynamic system. A step is divided into several stages. Each stage represents a typical interaction or activity among agents and spots in a given dynamic system. Each stage is divided into two phases called agent phase and resolver phase. In the agent phase, agent rules are executed at each agent turn. In the resolver phase a resolver on a spot calculates the interaction among agents and spots.

(2) Agent interaction by spots

Agent executes processing based on active role rule of agent itself, using simulation time, attribute information of agent itself, and attribute information of spots, and then updates attribute information of agent itself and attribute information of spots. Agent can not directly control attribute information of other agent, and indirectly interacts through the spots. The update of attribute information of agent itself is immediately reflected as well as normal variable. On the other hand, the update of attribute information of spots is resolved by the object called resolver, and updated result is reflected from next stage. The processing of this resolution is a primary difference between the traditional sequential processing SOARS and parallel SOARS.

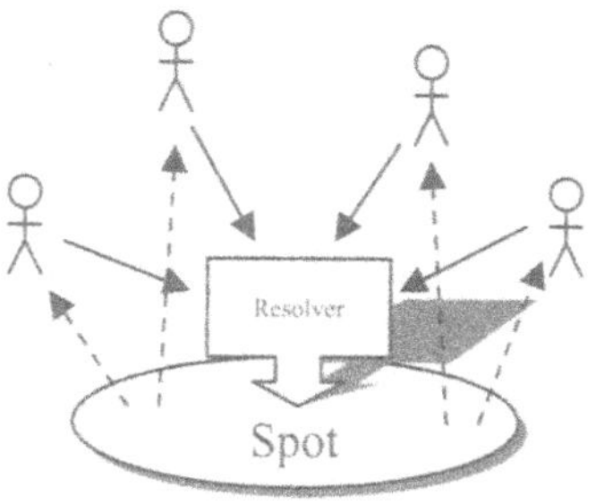

Fig. 1. Agents' Interaction by Spot Operation and Feedback

(3) Substitution by resolver

The conflict of spot attribute operation among multiple agents is resolved by object called resolver. Under the implementation of parallelizing SOARS, the order of agent's rule execution is no longer constant. In the case that the operation result of spot attribute fluctuates depending on the execution order of agent processing, simulation is in danger of instability without reproducibility. Therefore, parallelizing SOARS conciliates the operation against spot attribute with resolver, and guarantees updated result of spot attribute that does not depend on execution order. Since resolver also reduces restrictions of interlocking significantly, the effects to improve parallelism of rule processing can be expected. Substitution of spot attribute by resolver has the following variations:

– asserted substitution

When substitution of attribute value is processed by multiple agents under single stage at the same time, the last substituted value remains as updated value in traditional non-parallel substitution. On the other hand, in case of asserted substitution in parallelized SOARS, substitution is successful only when all substituted values are equal. When multiple conflicting substitutions are executed, an error occurs and simulation is interrupted.

The substitution result of attribute value is reflected on readout value from the next stage. That is, attribute readout value is the substitution result in the previous stage, and that makes it guaranteed that the equal value is always read out by all agents under the single stage. This stage delay of readout value is carried on in the case of substitution by other resolver.

– resolution by arithmetic operation

In number attribute such as integer number or real number, it is considered to process the result of arithmetic operation like sum of multiple substitution values as updated value. Regarding integer number and real number, examples of arithmetic operations are composed mainly of the followings:

- addition
- multiplication
- maximum
- minimum

Furthermore, regarding integer number, examples are logical operations like:

- bitwise and
- bitwise or
- bitwise xor

Above arithmetic operations are all associative[4] and commutative[5] binary operations. Mathematically defined, a combination of a set and a binary operator defined on the set which is associative and commutative is called commutative semigroup.

When associative and commutative binary operator on integer number or real number is indicated as $\circ$, and the substitution value of agent $i \in [1, n]$ is indi-

---

[4] If a binary operator $\circ$ fulfills the requirement $(a \circ b) \circ c = a \circ (b \circ c)$ for any elements a, b, c in a set, the operator $\circ$ is said associative.

[5] If a binary operator $\circ$ fulfills the requirement $a \circ b = b \circ a$ for any elements a, b in a set, the operator $\circ$ is said commutative.

cated as a_i, then attribute value after update v is indicated as the following formula:

$$v = a_1 \circ a_2 \circ \cdots \circ a_i \circ \cdots \circ a_n$$

Since the operator $\circ$ commutative in above formula, the same value v can be obtained no matter how terms on the right-hand side permute. For example, when addition + is used as operator, sum of $\{a_i\}$ can be obtained like below formula.

$$v = a_1 + a_2 + \cdots + a_i + \cdots + a_n = \sum_{i=1}^{n} a_i$$

This operation can be executed effectively under the implementation of multiprocessor. For example, we consider that processor A process agent of odd number index $(i = 1,3,\cdots)$, and processor B process agent of even number index $(i = 2,4,\cdots)$. Processor A and Processor B first calculate partial terms v_A, v_B under rule execution during the stage.

$$v_A = a_1 \circ a_3 \circ \cdots$$

$$v_B = a_2 \circ a_4 \circ \cdots$$

Then, after the stage, they execute operation among the partial terms, and update ultimate value.

$$v = v_A \circ v_B = (a_1 \circ a_3 \circ \cdots) \circ (a_2 \circ a_4 \circ \cdots)$$

Since the calculation of partial terms can be executed individually by each processor in above processing, it has no synchronism load and is very efficient.

– random select substitution

While the substitution of multiple different values in asserted substitution causes an error, random select substitution chooses one of all substitutions in the stage with equally probability and reflects as readout value on the next stage.

Also under multiprocessor implementation, random select substitution can be executed efficiently. For example, i times of substitutions are executed in random select processing with processor A as $a_1, a_2, \cdots, a_i$, and one substitution value among them selected equitably is defined as α. In the same way, j times of substitutions are executed in random select processing with processor B as $b_1, b_2, \cdots, b_j$, and one substitution value among them selected equitably is defined as β. In order to merge random select of processor A and B, we may select α with probability $i/(i+j)$ and β with probability $j/(i+j)$ and make it substitution value. In the same way, merging selection results of all processors gives us selection result with equally probability.

3 System architecture

(1) System components

The prototype system of parallel SOARS is composed of host PC and FPGA board. The FPGA board is connected to the host PC by some communication interface like RS232C, Ethernet, or PCI-bus. The FPGA board also equips some additional memory chips beside main FPGA chip, and usually available FPGA evaluation board has same constitution. We also use FPGA evaluation board on the market as the FPGA board for the prototype system of parallel SOARS.

In order to perform demonstration experiments of parallel SOARS, we developed multiprocessor core CPU with unique architecture refining its function as much as possible for inexpensive FPGA to get the most parallelism. We call this processor μSOARS since it is specialized in implementation of parallel SOARS.

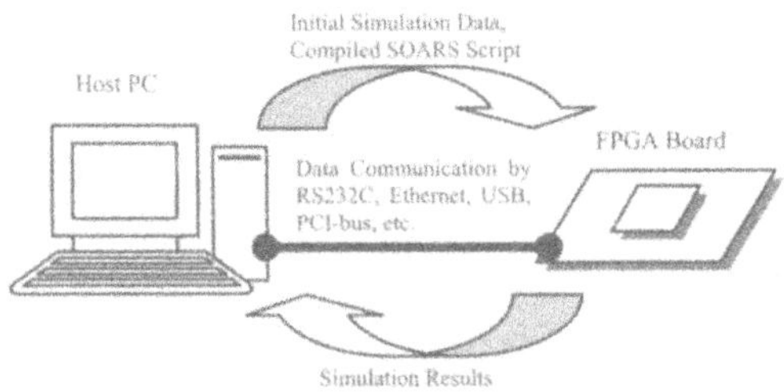

Fig. 2. System Components of Parallel SOARS

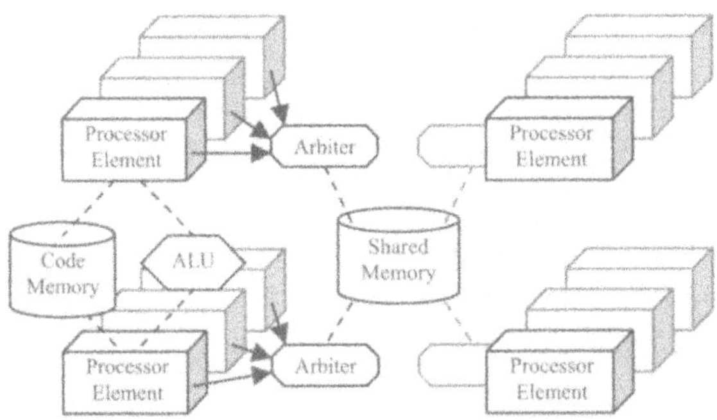

Fig. 3. Components of μSOARS Multiprocessor Core Architecture

(2) Agent rule processing sequence

Agent rule is basically processed sequentially assigning agent's processing loop to unoccupied CPU's on each stage by a method called work pile. Assignation of processing and collection of result is performed by a particular CPU, which is called controller CPU.

When simulation is started, attribute data of all agents is transferred from host PC to FPGA board and stored FIFO buffer consisting of external memory on the board.

During a stage, controller CPU sequentially transfers agents' attribute data from FIFO buffer to shared memory in FPGA, and continuously assigns agent rule execution processing to other CPU's one after another. CPU's assigned agent rule execution processing inform controller CPU that processing is completed. Controller CPU informed that processing is complete transfers agents' attribute data on shared memory to FIFO buffer and frees shared memory area.

At the end of the stage, controller CPU waits for all CPU's to complete execution, and then performs merge of contraction and update of spot attribute. When the step is over, update of simulation time and delivery of log information to host PC are executed, as well.

(3) Independent pseudorandom number sequence

The implementation of pseudorandom number generator affects efficiency and scalability of parallel processing for the simulation. Centralized random number generation may degrade scalability due to overhead of mutual locking for the generator. Processor-affiliated independent pseudorandom number generators may resolve the question, but the ordinary pseudorandom number generator does not provide the independency. Only changing random seeds is not sufficient, because when a random sequence of a generator accidentally catches up another sequence, then the simulation results in biased answer.

A related research of creating independent pseudorandom number generator is Dynamic Creator of Mersenne Twister[6]. We also attempt to provide simpler algorithm of independent pseudorandom number generator using LFSR (Linear Feedback Shift Register) extended by quadratic polynomials.

(4) Reproducible parallel processing

In ordinary non-parallel implementation, the simulation should reproduce fixed result for the same input and random seeds. This nature is important for the simulation to keep traceability. For high-performance parallel implementation, this property may break due to non-deterministic execution of agents' rule actions. At

---

[6] http://www.math.sci.hiroshima-u.ac.jp/~m-mat/MT/emt.html

small cost of performance, we can take back the reproducible nature for advanced implementation of parallel SOARS.

In SOARS modeling concept, the reproducibility may break at pseudorandom number generation or the operation of spot's attributes. Centralized or processor-affiliated pseudorandom number generators are not sufficient for the reproducibility. Agent-affiliated generators can provide trial-independent random sequences at the agents' rule actions and keep the reproducibility.

4 Conclusion

In the development of parallel SOARS, we realized that the consideration beyond the levels of abstract modeling concept and concrete hardware implementation is necessary to build the system. The social systems of real world are parallelized by nature, but the processing of ordinary computer systems are often confined within serial programming style. The collaboration of social science and computer science will lead us to new representation of social systems and new paradigm of parallel processing.

Acknowledgement

This research was supported in part by "The 21st Century Center of Excellence Program" of the Ministry of Education, Culture, Sports, Science and Technology, Japan and by Exploratory Software Project grant of Information-technology Promotion Agency, Japan.

References

[Deguchi 2004] Hiroshi Deguchi, Economics as an Agent Based Complex System, Springer-Verlag, 2004.

[Hennessy 1996] John L. Hennessy and David A. Patterson: Computer Architecture. A Quantitative Approach, Morgan Kaufmann, 1996, ISBN: 1-55860-329-8.

[Matsumoto 2000] Makoto Matsumoto and Takuji Nishimura, "Dynamic Creation of Pseudorandom Number Generators", Monte Carlo and Quasi-Monte Carlo Methods 1998, Springer, 2000, pp 56--69.

[Tanuma 2004] Hideki Tanuma, Hiroshi Deguchi and Tetsuo Shimizu, "SOARS: Spot Oriented Agent Role Simulator – Design and Implementation", Post-proceedings of AESCS'04, Springer-Verlag, 2004.

CPSIA information can be obtained
at www.ICGtesting.com
Printed in the USA
LVHW081822260420
654475LV00017B/1010